It should be [...] each generation to rediscover principles of process safety which the generation before discovered. We must learn from the experience of others rather than learn the hard way. We must pass on to the next generation a record of what we have learned

C. Ducommun

ring and a
ompany in 1961
ident American
.

he conceived,
he unique series
ety in the 1960s
complemented
Process Safety

safety

sharingtheexperience

improving the way lessons are learned
through people, process and technology

BP Process Safety Series

Safe Tank Farms and (Un)loading Operations

A collection of booklets describing hazards and how to manage them

This booklet is intended as a safety supplement to operator training courses, operating manuals, and operating procedures. It is provided to help the reader better understand the 'why' of safe operating practices and procedures in our plants. Important engineering design features are included. However, technical advances and other changes made after its publication, while generally not affecting principles, could affect some suggestions made herein. The reader is encouraged to examine such advances and changes when selecting and implementing practices and procedures at his/her facility.

While the information in this booklet is intended to increase the store-house of knowledge in safe operations, it is important for the reader to recognize that this material is generic in nature, that it is not unit specific, and, accordingly, that its contents may not be subject to literal application. Instead, as noted above, it is supplemental information for use in already established training programmes; and it should not be treated as a substitute for otherwise applicable operator training courses, operating manuals or operating procedures. The advice in this booklet is a matter of opinion only and should not be construed as a representation or statement of any kind as to the effect of following such advice and no responsibility for the use of it can be assumed by BP.

This disclaimer shall have effect only to the extent permitted by any applicable law.

Queries and suggestions regarding the technical content of this booklet should be addressed to Frédéric Gil, BP, Chertsey Road, Sunbury on Thames, TW16 7LN, UK. E-mail: gilf@bp.com

Published by
Institution of Chemical Engineers (IChemE)
Davis Building
165–189 Railway Terrace
Rugby, CV21 3HQ, UK

IChemE is a Registered Charity in England and Wales
Offices in Rugby (UK), London (UK), Melbourne (Australia) and Kuala Lumpar (Malaysia)

ISBN-13: 978 0 85295 531 4

First edition 2004; Second edition 2005; Third edition 2006; Fourth edition 2008

Typeset by Techset Composition Limited, Salisbury, UK
Printed by Henry Ling, Dorchester, UK

Foreword

The petrochemical industry is often perceived as only a processing business. The logistics aspects tend to attract far less focus, despite the fact that they often require far more time, energy and effort than the processing of products. As part of the activities that take oil from the drilling site to a retail station, ships, pipelines, truck, railcars or tanks are loaded or emptied every second.

It is therefore not surprising that an incident sometimes occurs in this huge logistical chain—a tank is overfilled, a truck is unloaded into the wrong tank, a pump leak, etc. This booklet is intended for those operators, engineers and technicians working on tank farms and (un)loading plants to make them aware of the possibility of common failures and their consequences, in order to adopt safe designs and practices to avoid the occurrence of such incidents.

I strongly recommend you take the time to read this book carefully. The usefulness of this booklet is not limited to operating people; there are many useful applications for the maintenance, design and construction of facilities.

Please feel free to share your experience with others since this is one of the most effective means of communicating lessons learned and avoiding safety incidents in the future.

JJ Gomez, Vice President Refining
Safety & Operational Excellence

Acknowledgements

The co-operation of the following in providing data and illustrations for this edition is gratefully acknowledged:

- BP Process Safety Network
- BP Refining Technology, Sunbury team
- BP Shipping, Sunbury team
- BP Whiting and Grangemouth Refineries
- BP Amsterdam Terminal
- Phil Myers, ChevronTexaco
- Dr. Neil Adams, International Refinery Services
- John R Lockwood, Clarence Liew, Shawn Hoh of ABS Consulting Singapore
- John Bond, IChemE Loss Prevention Panel member

Contents

Note: All units in this book are in US and SI systems

1

Atmospheric storage tanks

1.1 Introduction

Both raw materials and finished products have to be stored in tanks. The value of these stored materials and finished products is high. Any loss of product, whether it occurs through leaks, spills or fire can have severe financial consequences and a significant impact on the future of the business.

Situations, such as fires and explosions as shown in the photographs, are of particular concern as they can cause injuries and fatalities. Many such incidents have occurred in the industry and are likely to continue unless we learn the lessons from the past.

Spills and tank fires cause pollution of the environment (air, land and water), which is unacceptable. These incidents result in fines and clean-up costs, with the possibility that the local regulator may decide to withdraw the licence to operate as well as seriously damaging the industry's reputation.

A total of 28 major tank incidents were reported in an incident database of an international oil company over five years. The breakdown of causes is shown in the chart below.

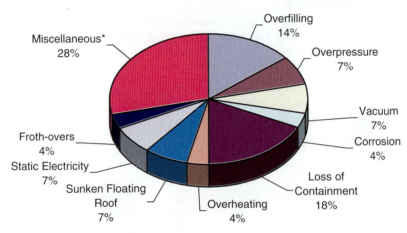

Breakdown of tank incidents by cause (1997–2001)

Overfilling 14%

Overpressure 7%

Vacuum 7%

Corrosion 4%

Loss of Containment 18%

Overheating 4%

Sunken Floating Roof 7%

Static Electricity 7%

Froth-overs 4%

Miscellaneous* 28%

*Miscellaneous causes include strong winds, lightning strikes, etc.

These tank farm incidents caused multiple fatalities and injuries. Nine resulted in fires. It becomes clear that although safety in the operation and maintenance of atmospheric storage tanks may not always be considered a high priority at some facilities, statistics show that tank incidents are frequent and can have catastrophic results.

This booklet is restricted to safe operations associated with non-refrigerated atmospheric tankage and the loading/discharging of railcars, road trucks and ships normally found at terminals. Readers should refer to BP Process Safety Booklet *Liquid Hydrocarbon Tank Fires: Prevention and Response* for matters related to fire protection and *Hazardous Substances in Refineries* for more detailed information and lessons learned on health and occupational hygiene matters.

For information on the safe handling of light ends and LPG, refer to BP Process Safety Booklet *Safe Handling of Light Ends*.

1.2 Types of tanks

In this booklet, the focus will primarily be on the following types of *low pressure atmospheric* storage tanks:

- cone roof/fixed roof tanks;
- floating roof tanks.

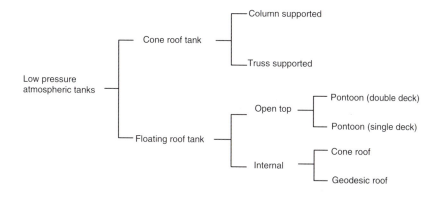

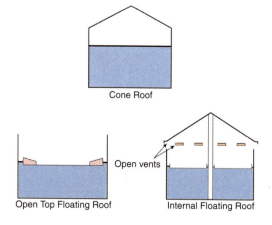

Cone Roof

Open Top Floating Roof   Open vents   Internal Floating Roof

Geodesic roof tank (Dome-shaped roof)

Some advantages are that the dome is supported only at the perimeter and requires no internal columns and supports. It does not need to be painted either on the inside or outside of the tank and it can be installed on existing tanks.

3

1.3 Fixed roof tanks

The two types of most commonly used storage tanks are:

- column supported cone roof, with a roof slope of 1:16 (gentle slope).
- truss supported cone roof, with a roof slope of 1:5 (steep slope).

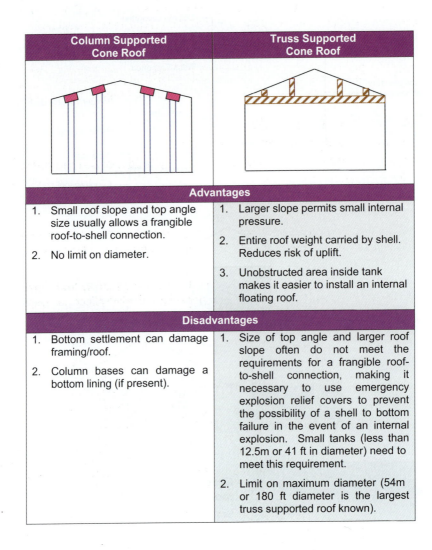

Column Supported Cone Roof	Truss Supported Cone Roof
Advantages	
1. Small roof slope and top angle size usually allows a frangible roof-to-shell connection. 2. No limit on diameter.	1. Larger slope permits small internal pressure. 2. Entire roof weight carried by shell. Reduces risk of uplift. 3. Unobstructed area inside tank makes it easier to install an internal floating roof.
Disadvantages	
1. Bottom settlement can damage framing/roof. 2. Column bases can damage a bottom lining (if present).	1. Size of top angle and larger roof slope often do not meet the requirements for a frangible roof-to-shell connection, making it necessary to use emergency explosion relief covers to prevent the possibility of a shell to bottom failure in the event of an internal explosion. Small tanks (less than 12.5m or 41 ft in diameter) need to meet this requirement. 2. Limit on maximum diameter (54m or 180 ft diameter is the largest truss supported roof known).

The air (vapour) above the liquid product in a fixed roof tank is in contact with the atmosphere through the breather vent. If the product is volatile enough, its vapour may mix with the air to form a flammable mixture. Flammable atmospheres should be avoided in fixed roof tanks through adequate control systems and when appropriate, nitrogen or fuel gas blanketing arrangements (steam may be used for bitumen tanks). Run down temperatures from process units should be kept below the flash point of the product—this is particularly important for products such as kerosene, where the ambient temperature may approach the flash point (see page 8 for definition). Also, tanks should be well protected from sources of ignition.

> **Flammable atmospheres should be avoided in fixed roof tanks.**

1.4 Floating roof tanks

The advantages of floating roofs are as follows:

- there is no vapour space, thus eliminating any possibility of a flammable atmosphere;
- reduction of evaporation losses;
- reduction in air pollution;
- internal floaters increase protection from fire exposure (therefore fitting a geodesic dome over a floating roof tank significantly decreases the probability of ignition).

Vapour emissions from internal floating roof tanks are generally lower than those from open top floating roof tanks because the wind effect has been eliminated.

Open top floating roof

The main types of floating roof tanks are presented below:

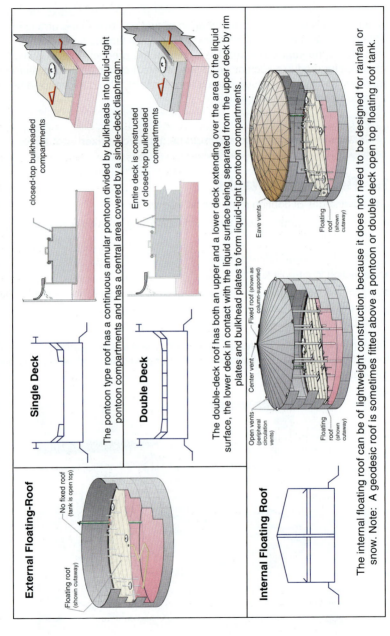

External Floating-Roof

No fixed roof (tank is open top)

Floating roof (shown cutaway)

Single Deck

closed-top bulkheaded compartments

The pontoon type roof has a continuous annular pontoon divided by bulkheads into liquid-tight pontoon compartments and has a central area covered by a single-deck diaphragm.

Double Deck

Entire deck is constructed of closed-top bulkheaded compartments

The double-deck roof has both an upper and a lower deck extending over the area of the liquid surface, the lower deck in contact with the liquid surface being separated from the upper deck by rim plates and bulkhead plates to form liquid-tight pontoon compartments.

Internal Floating Roof

Open vents (peripheral circulation vents)

Center vent

Fixed roof (shown as column-supported)

Floating roof (shown cutaway)

Eave vents

Floating roof (shown cutaway)

The internal floating roof can be of lightweight construction because it does not need to be designed for rainfall or snow. Note: A geodesic roof is sometimes fitted above a pontoon or double deck open top floating roof tank.

1.5 Different types of products in different tanks

The type of storage tank used for a specific product is principally determined by safety and environmental requirements and the need to operate economically when the tank is in service (minimize losses from evaporation). See page 8 for selection criteria.

> **The liquid's true vapour pressure is usually the main determining factor for the selection of type of storage tank.**

Difference between RVP and TVP

Reid Vapour Pressure (RVP) is determined at 37.8°C (100°F), in the presence of air, through standard laboratory procedures.

True Vapour Pressure (TVP) is the observed pressure, in absolute units, when a vapour is in equilibrium with its liquid at a constant temperature.

RVP and TVP are different because RVP is measured with the inclusion of air while TVP is gauged using an evacuated container. In effect, the Reid method replicates the conditions in which volatile petroleum products are stored and handled in contact with atmospheric air. RVP is about 6% less than TVP at 37.8°C (100°F).

Further, RVP is determined at a single temperature while TVP may be obtained for varying temperatures.

The Reid Method is one of the most common methods of measuring vapour pressures of petroleum fractions. It gives a measure of the inherent tendency of a product to evaporate.

Criteria for storing products in various tanks

Parameter	Criteria	Tank type	Comments
True Vapour Pressure (TVP)* —main determining factor	TVP < 1.5 psia (0.10 bara)	Fixed cone roof tank	Fixed roof tanks are generally used where the products stored do not readily vaporize at the ambient or stored temperature conditions.
	1.5 ≤ TVP ≤ 11.1 psia (0.10 ≤ TVP ≤ 0.77 bara)	Floating roof tank	US New Source Performance Standards ? EPA & BP Standard
	TVP > 11.1 psia (TVP > 0.77 bara)	Low pressure storage tank or pressure vessel	Product is too volatile (gaseous) for floating roof tank (roof may collapse due to instability). Tank emissions should be recovered for recycle or destruction.
Flash Point (FP)**	FP > 55°C (130°F)	Fixed cone roof tank	For example, gas oils, diesel oils, fuel oils.
	Low FP, <55°C (130°F)	Floating roof tank	For example, gasoline, crude oil, naphtha. (Liquids that could produce a flammable atmosphere at normal outside temperatures are normally stored in floating roof tanks, except when the vapour space is kept inert (free of air) by a nitrogen blanket and protected by pressure/vacuum valves.)

* *True Vapour Pressure (TVP) is the pressure, in absolute units, exerted by a vapour that is in equilibrium with a liquid at a given temperature. When the TVP of a product is high, evaporation becomes excessive. If exposed to atmosphere, there could be product loss and a flammable cloud may be formed.*

** *The Flash Point (FP) of a liquid is the lowest temperature at which it gives off enough vapour to form an ignitable mixture with air.*

Introducing products with a too high TVP or increasing the TVP of products by injecting light ends in a tank is a hazardous activity, as shown by the incidents described below.

ACCIDENT Batch blending was going on in a 7,000 m³ unleaded gasoline tank when a fire occurred. During more than 30 hours, 56 fire trucks tried to tackle the fire, successfully protecting adjacent tanks. The investigation team found that the blending calculations were wrong and three times too much butane was being sent to the tank. A bubble of light ends probably lifted and tilted the roof, creating enough static or metal to metal friction to ignite the vapours.

The blending of butane into gasoline base stock in atmospheric storage must be avoided. A closed circulating pipe system (under pressure) should be used. When butane is introduced into a tank containing a low vapour pressure stock, the initial charge of butane may cause vapour agitation of the stock, especially if the liquid level is low. Such agitation may generate electrostatic sparks on the oil surface and result in a serious explosion.

Bad control of blending operations has caused multiple floating roof sinkings (see the BP Process Safety Booklet *Safe Handling of Light Ends* in this series).

The incident described in section 7.5.9. was caused by introducing too high TVP Naphtha in a floating roof tank.

1.6 Classification of liquids

Classification (NFPA)	Class	Flash Point	Boiling Point	Preferred type of tank
Flammable liquid (A liquid having a closed cup flash point below 100°F (37.8°C) and having a vapour pressure not exceeding 40 psia (2068 mmHg) at 100°F (37.8°C) shall be known as a Class I liquid)	1A	<73°F (22.8°C)	<100°F (37.8°C)	Floating roof
	1B	<73°F (22.8°C)	>100°F (37.8°C)	
	1C	73°F ≤ X < 100°F (22.8°C≤X<37.8°C)	N.A.	
Combustible liquid (A liquid having a closed cup flash point at or above 100°F (37.8°C))	2	100°F ≤ X < 140°F (37.8°C ≤ X < 60°C)	N.A.	
	3A	140°F ≤ X < 200°F (60°C ≤ X < 93°C)	N.A.	Fixed roof
	3B	≥200°F (93°C)	N.A.	

N.A. = Not Applicable

European Model Code of Safe Practice

Class of Liquids	Flash Point	Storage condition	Preferred type of tank
I	<21°C (<69.8°F)	–	Floating roof
II (2)	21–55°C (69.8–131°F)	Handled at a temperature above its flash point*	
III (1)	55–100°C (131–212°F)	Handled at a temperature below its flash point*	Fixed roof
Unclassified	>100°C (>212°F)	–	

* Dependent upon ambient and process rundown temperatures.

1.7 Flammable limits

Lower flammability limit/lower explosive limit (LFL/LEL) corresponds to the minimum concentration of vapour in air necessary for combustion.

Upper flammability limit/upper explosive limit (UFL/UEL) is the maximum concentration of combustible vapour in air which will burn.

> **The flammable range is the concentration range lying between the lower and upper flammability limits (LFL/LEL and UFL/UEL).**

The terms LFL and LEL, and UFL and UEL are interchangeable.

Flammable vapour/mist will only ignite if the mixture is within the upper and lower flammable limits.

9

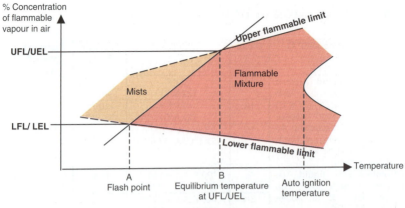

Concentration against temperature

Mists can form flammable mixtures below their flashpoints.

The upper flammable limit concentration generally increases with temperature and the lower flammable limit decreases. Thus the range broadens as the temperature increases.

Auto ignition temperature is the lowest temperature at which a liquid will ignite without the direct application of flame.

The diagram below illustrates the flammable range for gasoline vapour in air.

The following table shows the flash points and flammable limits of some common petroleum and chemical products:

Product	Flash Point		LFL	UFL
	(°C)	(°F)		
Gasoline	-45	-49	1.4	7.6
Hexane	-22	-8	1.2	7.4
Xylene	29	84	1.0	7.0
Styrene monomer	31	88	1.1	6.1
Fuel Oil No 1 (Kerosene)	43–72	109–162	0.7	5.0

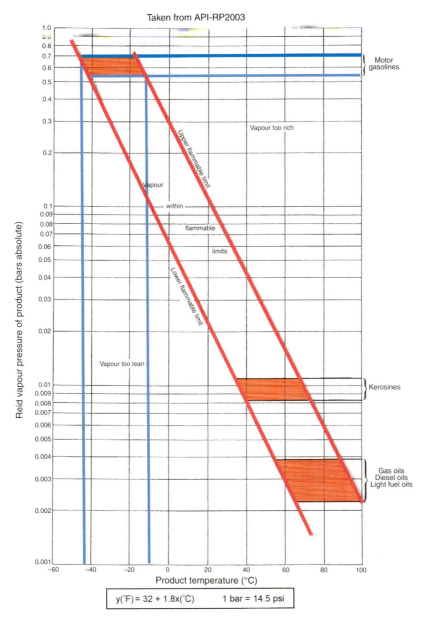

Taken from API-RP2003

Reid Vapour Pressure against Temperature showing the flammable range for various petroleum products

Other ways in which flammable atmospheres and explosions can be generated in fixed roof tanks are:

- Degradation of fuel oils in storage, particularly those containing cracked or visbroken residues. Despite quenching and stripping facilities on the process units, cracking reactions can continue in downstream piping, with resulting gas coming out of solution in storage tanks. Degradation can also occur in heavy product tanks contaminated with light ends.

- It is very easy to produce a flammable atmosphere above a low volatile product like fuel oil/gas oil through blending, contamination or from overheating the contents. For example, as little as a 3% blend of gasoline to gas oil brings the flash point down from 70°C (156°F) to 10°C (50°F). A heavy oil contaminated in this way can give off sufficient vapour to provide a flammable atmosphere. Such contamination can happen where common pipelines/manifolds are used for high and low volatile fuel.

- Poor segregation of light and heavy slops systems.

- Heating tanks containing light slops.

- Exceeding maximum heating temperature and minimum tank dips (exposing heating coils) for heavy slops tanks.

- Splash filling and jet mixing creating mists with an inherent source of ignition (by static generation).

- Mixing incompatible waste materials, such as pumping an oxidizing mineral acid into a tank containing hydrocarbon material or mixing sulphuric acid with ferric chloride.

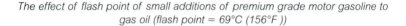

The effect of flash point of small additions of premium grade motor gasoline to gas oil (flash point = 69°C (156°F))

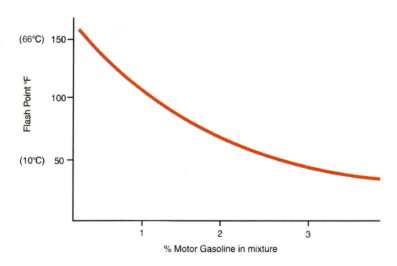

ACCIDENT Lightning ignites flammable atmosphere within a fuel oil tank!!!

An explosion and a fire occurred when lightning struck a fuel oil tank. The investigation showed that the fuel oil contained enough propane to create a flammable atmosphere below the roof. The fuel oil was a product stream from a propane deasphalting unit.

2

Important design issues

2.1 How strong is a storage tank?

Large fixed roof tanks are vulnerable due to their low resistance to internal/external pressure.

Floating roofs are vulnerable to sinking due to loss of buoyancy through leaking pontoons or excessive accumulation of water.

Fixed roof tanks may:
- withstand overpressure of 7.5 mbar (3 inches H_2O).
- withstand vacuum of 2.5 mbar (1 inch H_2O).
- have weak shell-to-roof welds that open (fish mouth) to minimize damage in the event of overpressure.

Floating roof tanks may have buoyancy to remain afloat with the primary drains inoperative for the following conditions:
- 10-inch water over a 24-hour period with roof intact, except for double deck roofs equipped with emergency drains.
- any two compartments punctured.

3" water gauge is the pressure at the bottom of a cup of tea (7.5 mbar, 1/10 psi)

1" water gauge is the pressure at the bottom of a perfume bottle (2.5 mbar, 1/30 psi)

Storage tanks always look big and strong, but compare the strength of a Heinz Baked Bean Can with that of a storage tank with regard to internal pressure. A small pressure over a large area makes a big force. Refer to BP Process Safety Booklet *Hazards of Trapped Pressure and Vacuum.*

If a Heinz Baked Bean Can has a strength = 1, then,

Item	Shell	Roof
Heinz baked bean can (small)	1	1
40 gallon drum (151 litres)	1/2	1/3
1,800 ft³ tank (13,465 gallons/50,971 litres)	1/3	1/8
3,600 ft³ tank (26,930 gallons/101,941 litres)	1/4	1/11
18,000 ft³ tank (134,650 gallons/ 509,706 litres)	1/6	1/33
36,000 ft³ tank (269,300 gallons/1,019,410 litres)	1/8	1/57

The next time you eat a can of baked beans, just see how easy it is to push the sides or top in with your fingers—and then look at the table again.

The bigger the fixed/cone roof storage tank, the more vulnerable it is to failure from overpressure!

2.2 Tank overpressure and vacuum

Operations and situations which can lead to overpressure or vacuum in an atmospheric storage tank.

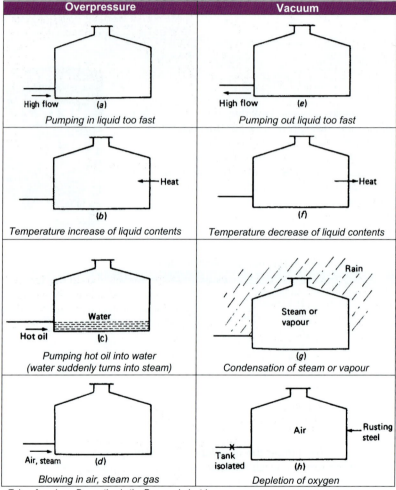

Overpressure	Vacuum
High flow (a) — *Pumping in liquid too fast*	High flow (e) — *Pumping out liquid too fast*
Heat (b) — *Temperature increase of liquid contents*	Heat (f) — *Temperature decrease of liquid contents*
Water / Hot oil (c) — *Pumping hot oil into water (water suddenly turns into steam)*	Rain / Steam or vapour (g) — *Condensation of steam or vapour*
Air, steam (d) — *Blowing in air, steam or gas*	Air / Rusting steel / Tank isolated (h) — *Depletion of oxygen*

Taken from Loss Prevention in the Process Industries

Overpressure or vacuum can lead to the bursting or collapse of an atmospheric fixed roof storage tank.

The shell to roof welded seam should always fail first—this is called a 'frangible' weld. For tanks less than 15m (50 feet) the roof may not be frangible—liquid can be released during overpressure by separating the shell to bottom joint. If the roof is not frangible, ensure that emergency venting is provided especially for small tanks.

Shell to roof weld rupture

Shell to bottom weld rupture (tank overpressured when purging a line with nitrogen)

Roof plates are not attached to the supporting structure but to the top curb angle of the tank shell only. The purpose of a frangible joint is to allow relief of an abnormally high overpressure through a failure of the weak roof-to-shell connection and not via a ruptured tank shell or bottom-to-shell connection, safely containing the product inside the tank.

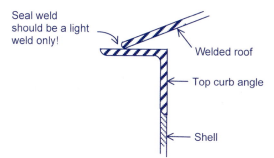

Seal weld should be a light weld only!

Welded roof

Top curb angle

Shell

Note that geodesic dome roofs are not frangible.

ACCIDENT Small tank rockets into the sky!!!

This 30.5m (100 ft) high, 6.1m (20 ft) diameter storage tank exploded and rocketed 91.4m (300 ft) into the air when exposed to fire initiated by a road truck hitting a nearby loading gantry.

The value of fragile roof-to-shell seams or emergency reliefs to avoid bottom-to-shell failure is clearly demonstrated by this incident. It is not always practicable to have weak roof-to-shell seams for small diameter tanks.

Fixed roof tanks are equipped with Pressure Vacuum Valves (PVVo) or vents to allow air ingrooo and egress so that a constant pressure can be maintained inside the tank. Roof vents function to allow breathing and filling losses that can cause overpressure and vacuum in fixed roof storage tanks (i.e. tanks with a vapour space above stored liquid).

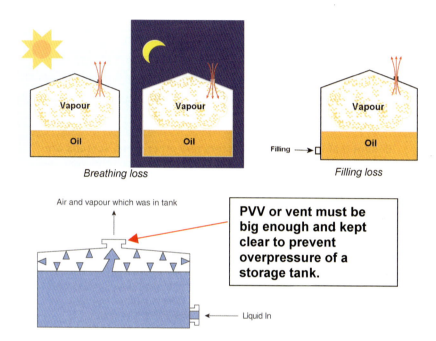

Breathing loss

Filling loss

Air and vapour which was in tank

PVV or vent must be big enough and kept clear to prevent overpressure of a storage tank.

Liquid In

- For liquid to get in, air and vapour must be pushed out.
- The pressure in the tank must be slightly above atmospheric pressure.
- The tank should be designed to withstand a pressure of **7.5 mbar** (**3″ water**).

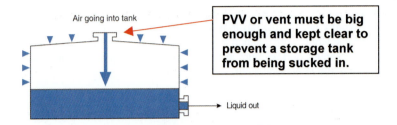

Air going into tank

PVV or vent must be big enough and kept clear to prevent a storage tank from being sucked in.

Liquid out

- For liquid to get out, air and vapour must be sucked in.
- The pressure in the tank must be slightly below atmospheric pressure.
- The tank should be designed to withstand a vacuum of **2.5 mbar** (**1″ water**).

These requirements are incorporated in **API 2000** for tanks designed to API Standards.

> **Ensure that vents, flame arrestors and other mesh screens are regularly inspected to ensure that they are clear of rags, rust, ice, tissues, wax, plastic, polymer and other debris.**

ACCIDENT Blocked vent causes partial tank collapse!!!

A 90,000 tonnes jet fuel tank partially collapsed some 30 seconds after product withdrawal had stopped. The collapse was due to a blockage in the vent system, which had caused a slight internal vacuum. The 200 mesh gauze, which had not been checked for nearly a year, was choked with sand and corrosion products. Fortunately, there was no other damage and the buckled plates sprang back when the tank was filled with water.

Rust partially blocks a screen mesh on a fixed roof tank's open vent

The use of fine gauze flame arrestors is not recommended without regular inspection for tanks because of their tendency to block. A coarse gauze may be fitted if required to keep out debris and prevent birds nesting. Always make sure the vent or vacuum/pressure valve is **clear** and that the venting system is never modified without authorization through a management of change procedure.

Some vent conditions which can lead to the bursting or collapse of an atmospheric storage tank

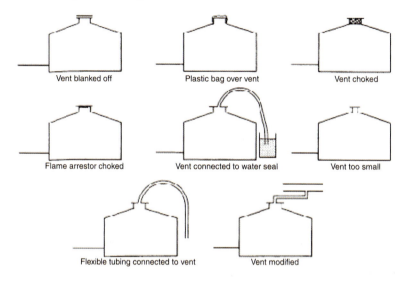

Vent blanked off	Plastic bag over vent	Vent choked
Flame arrestor choked	Vent connected to water seal	Vent too small
Flexible tubing connected to vent	Vent modified	

Taken from Loss Prevention in the Process Industries

> **Refer to *API Standard 2000* for further details on causes of overpressure and vacuum, venting requirements, and installation and maintenance of venting devices.**

2.3 Pressure/vacuum valve

Pressure/vacuum valves are normally fitted to cone roof tanks which contain more volatile products in place of open vents. They are designed to protect the tank from vacuum or over-pressure. They are also designed so that an external source of ignition cannot propagate into the tank headspace during filling or emptying.

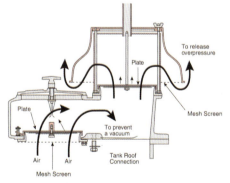

While these PV valves do not act as flame arrestors, it has been shown that their mode of operation effectively prevents flame propagation into the tank provided the valve is not connected into a piping system. A flame arrestor is therefore not considered necessary for use in conjunction with a PV valve which is directly venting to atmosphere.

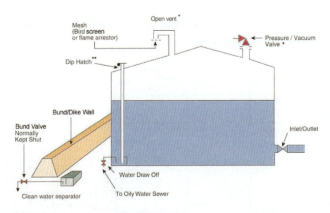

Typical arrangement for a fixed roof tank

Notes * A tank is either fitted with an open vent or pressure/vacuum valve dependent upon the stored liquid's classification (see *Section 1.6: Classification of liquids*). Open vents are installed on tanks containing low volatile and high flash point liquids. Pressure/vacuum valves are installed on tanks equipped with a gas blanket or if the tank could contain high volatile or low flash point liquids.

** Warning! Opening these dip-hatches when a tank is fitted with a PV valve will release a significant quantity of vapour.

> **Protect against the inhalation of vapour which escapes as a momentary 'whoosh' when the gauge hatch is opened on a fixed/cone roof tank equipped with a PV valve.**

2.4 Location of storage tanks

The arrangement and layout of storage tanks should take into account:

- normal operation;
- emergency operation;
- firefighting activities.

The design of a tank farm should take into account the likely consequences of any accidental spillage or fire. Storage tanks should be located away from potential sources of ignition, and spacing provided to minimize the possibility of knock-on radiation from any fire, which could possibly occur in an adjacent area.

NFPA 30 standard on tank location and bunding/diking should be regarded as a strict minimum. Spacing is often the only good, fail-safe measure to prevent escalation in case of fire. Refer to the BP Process Safety Booklet *Liquid Hydrocarbon Storage Tank Fires: Prevention and Response* for more details.

> **Locate storage tanks away from potential sources of ignition.**

Roads around tank farms and separating tank plots are required for a variety of activities such as tank cleaning, maintenance and repair as well as to enable rapid response to emergency situations. Major roads should be free of access restrictions. Minor roads, for maintenance purposes, may pass through hazardous areas but will need barriers to restrict or control access.

Good access to tanks for firefighting purposes requires a limit of two rows of tanks between roads. This is especially important with large tanks where ease of access to all sides in the early stages of the fire may be critical.

Typical floating roof tank farm with each large tank having its own individual bund/dike.

Example of a crude floating roof tank contained in metallic bund.

23

ACCIDENT **Poor location of storage tanks and poor access: Tacoa power station boilover, Venezuela, 19 December 1982**

On 19 December 1982, in a power plant in Tacao near Caracas (Venezuela), a huge boilover (see explanation below) occurred on a fuel oil tank, killing at least 160 people in a huge fire ball. The installation comprised three power plants (one under construction) by the sea side, and a tank farm on the hill above. The site was surrounded by a poor residential area.

On 18 December, operators transferred fuel oil #6 from tank 9 to tank 8, on top of the hill (see drawing below giving approximate distances and elevations). Tank 8 was a 40,000 m³ fixed roof tank of 55 m (180 ft) diameter, filled approximately with 14,000 m³ of fuel oil.

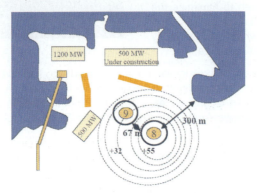

At 23:30, seeing that the product temperature was way too high (88°C/190°F instead of 65°C), operators cut a tracing system to allow temperature to cool off. The next morning before dawn (around 06:00), two operators went to manually gauge the tank. During the operation, an explosion occurred, either due to the use of non-intrinsically safe lamps or to a static spark. The temperature of the product was still above 80°C/176°F, well above the flash point of 71°C/160°F. The roof of the tank landed some distance away and severed some product lines, igniting a bund fire. The roof also supported the foam systems which were ripped off when the roof took off. Tank 8 immediately suffered a full surface fire.

Fire fighting action was limited by the remoteness of the site, inadequate access to the tank (hilly terrain and the only access road was below the bund on fire) and damage caused to the fixed fire protection systems. Fire fighters, civil defence personnel, plant workers, journalists and onlookers were within 30 to 60 metres (100 to 200 ft) of the tank.

At 12:20, a massive boilover occurred, with a fire ball approximately 150 m (500 ft) diameter, raising up to 600 m (2,000 ft) high, and burning product fell down around the site.

continued

The boilover pushed product over the top of the tank, creating a wave of burning liquid that went over the 6 m (20 ft) high bund wall and submerged vehicles and people alike and travelled more than 400 m (1,300 ft) downhill. This wave entered another bund and ignited another fuel oil tank, which went on to burn for several days. Tank 8 fire was extinguished by the boilover.

This incident killed at least 160 people including 40 firefighters and eight journalists, injured more than 500, destroyed 60 vehicles and 70 houses and most of the power generation plant (1200 MW and one 500 MW section destroyed). 40,000 people were evacuated by the army as another boilover was feared from the other tank on fire.

Before the fire, consideration was not given to the potential of a tank farm major leak impact on the power plants. Refiners have learned at their expense never to locate tanks above process units—even if tank incidents are rare events, the inventory located in each single vessel is much greater than in process units. Plant layout is an essential part of the plant inherent protection—a good layout can help firefighters to have good and safe access to an emergency and keep the necessary water resources to a strict minimum by having good separation distances. A poor layout may mean that a whole plant will be lost in a single event that could have been controlled otherwise.

To read more on boilovers, refer to BP Process Safety Booklet *Liquid Hydrocarbon Tank Fires: Prevention and Response.*

Bunding/Diking	
Bund/dike walls	• Tanks should be surrounded by a bund/dike wall capable of containing at least the capacity of the largest tank within the bund/dike. • The walls and floor of the bund/dike should be impervious to liquid and designed to withstand a full hydrostatic head. • Bund/dike walls should be below 2m (6.6 ft) to provide adequate natural ventilation, ready access for firefighting, and good means of escape in any emergency situation. • Intermediate walls of less than 0.5m (1.6 ft) height can be used as a convenient way to contain small spillages and act as firebreaks. • The floor of the bund/dike should be sloped to prevent minor spillages remaining around the tank. • The distance between tank and bund/dike walls must be at least **1m (3.3 ft)**. • LEL detection should be installed in all bunds containing low flash point products (gasoline, methanol, etc.).
Tanks	• All tanks should be sited on an impervious base. • The number of tanks in a bunded/diked compound should not exceed six or the total capacity exceed 60,000 m³ (15.9 million US gal). • Small tanks (less than 10m diameter) may be sited together in groups with total capacity below 8,000 m³ (2.1 million US gal). Such a group is regarded as one tank. • An individual bunded/diked compound should be provided for each tank above 30,000 m³/190,000 barrels (48 m/157 ft diameter). • It is recommended that the tank be located excentrically inside the bund/dike in order to improve the accessibility of the tank in case of a fire. • Do not store compressed, liquefied and pressure dissolved gases in a bunded/dike compound with other petroleum products. • No combustible materials or equipment should be stored in the bund/dike or against the bund/dike wall.
Drainage system	• The bund/dike drainage system should ensure that water discharge is controlled by a valve or a suitable pump on the outside of the bund/dike. • Liquid hydrocarbons automatic detection is a good practice. • If valves are used, procedures should be in place to ensure they remain closed, except when water is being drained off.

ACCIDENT **Small tank is washed away in a flood!!!**

Small tanks should be fixed to the ground in case of flooding. Ensure that procedures to inspect drain/bund/dikes during heavy rains are followed.

Early detection of leaks is essential to prevent large contamination of soil or ignition of a large vapour cloud if the leaking product has a low flash point. The incident described in section 6.1 is an example of what can happen if the incident is not detected at an advanced stage.

Example of an automatic liquid hydrocarbon detector (the four balls will sink deeper in oil than on water and this will close a circuit, triggering an alarm) associated with a LEL detector (black box).

- **Bund/dike should be capable of containing at least the capacity of the largest tank within the bund/dike.**
- **Bund/dike drain valves must be kept closed except when the bund/dike is being drained of rainwater.**
- **Incompatible chemicals should not be stored in tanks within the same bund/dike.**

NFPA 30 standard on tank location and bunding/diking should be regarded as a strict minimum. Spacing is often the only good, fail-safe measure to prevent escalation in case of fire. Refer to the BP Booklet *Liquid Hydrocarbon Storage Tank Fires Prevention and Response* for more details.

2.5 Rainwater on tank roof

Rain falling onto a floating roof must be drained away to the outside of the tank. Water is directed from the roof via a flexible hose, or a metal pipe with articulated joints, to an outside gate valve near the bottom of the shell. These valves should preferably be kept *OPEN*. If the policy is to keep them closed, this must be clearly stipulated as they will need to be opened in heavy rainfall. Tank roofs should be kept clear of debris to allow rainwater to flow into the drain collection sumps.

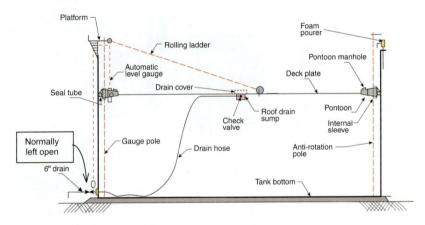

Leakage of tank contents into the roof drain system and bund/dike can occur due to hose failure. When this occurs, operators will keep the roof drain valve at the tank shell closed. However, this increases the possibility of a roof sinking due to product/water accumulation on the roof either due to the failure of the check valve to prevent backflow, or heavy rainfall.

Although there is a risk of the tank contents emptying into the dike/bund should a failure occur in the internal hose or metal articulated pipe, the likelihood of failure is lower than the chances of someone forgetting to open the valve during heavy rain that could result in the sinking of the floating roof.

> **Valves on roof drains should be kept OPEN at all times.**

Open top floating roofs should be inspected after heavy rain to ensure that drains are flowing and that standing water has not caused the roof to tilt or hang up.

Remember also that water freezes in cold winter areas. Follow the winterization instructions before cold weather, otherwise rupture of the drainage system may occur.

> **A tank with a leaking internal drain should be taken out of service as soon as practicable for repair.**

ACCIDENT

Rust and debris on floating roof can block the internal drain

Sunken floating roof due to accumulation of rainwater on the roof

An open top floating roof is designed to carry at least a load of 10 inches of rainwater before it will sink or to withstand deck (single pontoon) and any two compartments punctured (API 650). Sometimes less rainwater may cause the roof to sink particularly if any of the pontoons are damaged or leaking.

ACCIDENT Heavy rain produced flood conditions at a pipeline station, producing a large accumulation of water on floating roof tanks and flooding of the tank farm, switch gear and support buildings. The flooded tank dikes resulted in limited access to tank roof drains. Excessive water on the roof of a 150 foot (46 m) diameter double deck external floating roof tank caused the roof to sink. The exposed tank contents, 140,000 bbl (22,300 m^3) of a crude naphtha mixture, produced objectionable nuisance odours resulting in numerous complaints from local communities. Specialized foam equipment and personnel were used to maintain a foam blanket on the exposed tank contents until the product could be safely removed from the tank (refer to BP Process Safety Booklet *Liquid Hydrocarbon Tank Fires: Prevention and Response* for more details on the hazards of foaming operations).

Roof sunk　　　　　　　　　*Flooded tank farm*

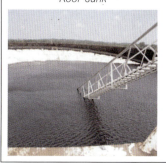

The picture below shows an empty tank that was lifted and floated away from its foundations after a hurricane flooded the bunded area.

Internal floating roof

Internal floating roof tanks can have any kind of floating roof, steel, aluminium, plastic. These roofs typically do not have closed drain systems since they are not exposed to rain. Often they will have heavy steel annular pontoon or double deck roofs when the tank was once an external floating roof but was converted to an internal floating roof tank (tanks with geodesic domes often have internal steel double decks or annular pontoons when they are storing volatile hydrocarbon liquids such as gasoline since they were simply converted from an external floating roof to an internal floating roof by covering the tank with a dome. However, domes can cover tanks storing diesel or other liquids in which case there may not be an internal floating roof. Tanks like this are simply fixed roof tanks).

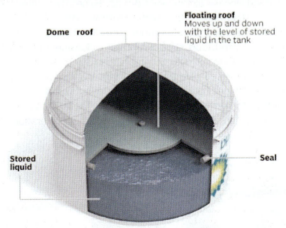

Dome roof

Floating roof
Moves up and down with the level of stored liquid in the tank

Stored liquid

Seal

An internal type roof (inside a freely ventilated cone/ fixed roof) is not designed for any load of product, foam or rainwater (see example of incident below and in Section 8). It does not have a drain and is less stable than an open top floating roof. The cone/fixed roof must be kept in good condition to prevent any possible ingress of rainwater.

ACCIDENT An incident occurred when firefighters tested foam systems on a tank fitted with an internal floating roof. The tank had been emptied, cleaned and inspected and a check of the firefighting systems was due before it could be put back in service. As the internal roof was not designed to withstand the weight of water, it collapsed when foam was applied.

Collapsed Internal Roof

A pan type roof (as shown below) has no closed buoyancy compartments, so it does not qualify as a floating roof for the siting requirements of NFPA 30 and is not recommended on any service as a single small leak will be enough to sink it.

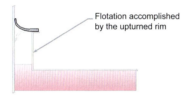

Flotation accomplished by the upturned rim

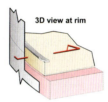

3D view at rim

An improved version of the pan roof is the bulkheaded pan which has open-top compartments. This is also not recommended as it is not inherently buoyant. During a firefighting exercise, the pontoons are all open and fire water will easily sink this roof. Also, if several compartments become filled, there can be a cascading failure of the pontoons where liquid overtops each compartment sequentially (the 'Titanic syndrome').

Aluminum roofs are cheaper than steel ones but they tend to get pinhole corrosion even in gasoline or other hydrocarbon type service. This leads to leaks which reduce bouyancy. In addition, there are hazards to personnel when performing maintenance work. Therefore, steel may be preferred.

A study carried out by BP in the 1980s has demonstrated that the atmosphere above floating roofs in well maintained gasoline tanks fitted with geodesic domes never reaches the Lower Flammable Level (in fact, the highest reading over 18 months was less than 0.5% of LFL). Recent updates of the LASTFIRE study, being carried out as a Joint Industry Project, confirm that fixed roof tanks fitted with an internal floating deck, or floating roof tanks fitted with a geodesic dome have a very low probability of suffering an internal fire.

In most known cases where such tanks were involved in a fire, the cause was linked to either:

- an external ignition (i.e. a tank overfilled and fire came back to the tank after the spill was ignited);

31

- a flammable atmosphere was allowed to build-up below the floating roof/deck (either during operations by draining too much product such as the 2nd incident described in section 3.1.5, or just before or after maintenance such as in the last incident described in detail in section 8).

Water explosions/frothover

It is quite common to find water in the bottom of storage tanks. Water in the bottom of a tank of hot oil is a potentially serious hazard. Even when the tank is normally operated well below the temperature at which water turns to steam, there is always the possibility of accidental overheating through failure of temperature controls or insufficient cooling of the feed before it enters the tank.

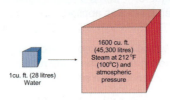

1cu. ft. (28 litres) Water

1600 cu. ft. (45,300 litres) Steam at 212 °F (100°C) and atmospheric pressure

The sudden formation of large quantities of steam causes a pressure surge that can overpressure the tank. At atmospheric pressure, a volume of water expands about 1,600 times when it flashes to steam. Violent foaming action caused by the vaporization of the water will result in a 'frothover'. The tank normally fails at the roof/shell seam if it is overpressured.

The viscosity of some heavy products, such as asphalt components may make higher storage temperatures, even above 150°C (300°F), necessary. These tanks must receive special attention and every precaution must be taken to prevent any water contamination.

> **Violent foaming action caused by the vaporization of water will result in a 'frothover'**

Foaming in an atmospheric storage tank

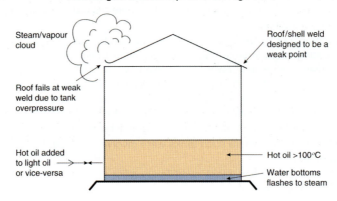

Steam/vapour cloud

Roof/shell weld designed to be a weak point

Roof fails at weak weld due to tank overpressure

Hot oil added to light oil or vice-versa

Hot oil >100°C

Water bottoms flashes to steam

ACCIDENT **Frothover destroys atmospheric residue tank!!!**

The photograph shows an incident which occurred when a 15 barg (220 psig) steam heating system was mistakenly left running for several days on an atmospheric residue tank containing water (as is often the case with product received from ships). When the temperature was high enough to vaporize the trapped water, a frothover occurred, damaging the tank beyond repair. Hot product was also projected over a large area. This could have resulted in a bad fire, had an ignition source been found.

Sudden mixing of products of different vapour pressures at different temperatures can also cause rapid evolution of vapour, or foaming, in a tank. This can happen when:

- hot oil is added to a tank containing a high vapour pressure product;
- a high vapour pressure product enters a hot tank;
- a heating coil disturbs stratified layers of such products;
- the breakdown of an emulsion disturbs stratified layers of such products.

Precautions to prevent frothovers

- If there is a possibility of water being present in a tank, the contents must be kept below the boiling point of water (not exceeding 93°C (200°F)) to prevent flashing to steam causing a violent eruption within the tank.
- Water should be drained frequently from the tanks.
- Wherever possible, tanks should be operated and maintained either well below the boiling point of water so that water bottoms will not flash to steam, or they should be kept sufficiently hot at all times so that water bottoms cannot accumulate.
- Tanks should not be operated in a range where temperatures fluctuate above and below the boiling point of water.
- Avoid accidental injection of water or light hydrocarbons where storage temperatures could exceed their boiling point.
- Provide run-down lines and tank-heater temperature controls for heated tanks. Tank storage temperatures must be carefully controlled. Flammable vapours can be generated if the temperature approaches the flashpoint of the product.
- Steam supply lines to tank heaters should be equipped with positive shutoff valves that will automatically close when the tank temperature reaches a predetermined set point.

Slop oil tanks are particularly susceptible to frothovers due to the nature of their operations. They receive oil-water mixtures from a variety of sources and function as separators, where settling produces oil streams with very low water content. The presence of water in the tank and varied temperatures and compositions of incoming streams have caused many frothover incidents.

ACCIDENT **Vapour cloud explosion!!!**

A vapour cloud occurred at a refinery in 1968 when heating caused a sudden vapour evolution in a slops tank and produced an explosion equivalent to 20 tons of TNT. The explosion caused the death of two people and injured 85 (mainly due to flying glass). The resultant fire covered an area of about 250m × 300m (820 ft × 984 ft).

The tank was full of mixed slops that had accumulated from many process units. Steam had been supplied to the tank's heating coils and it is believed that a water/oil emulsion separated at 100°C (212°F) suddenly releasing boiling lighter hydrocarbons. It was estimated that 50–100 tonnes of hydrocarbon vapours were released.

ACCIDENT **Incorrectly lined valves resulted in frothover!!!**

Hot vacuum gas oil at 204°C (400°F) was accidentally transferred to a cold slops tank containing a mixture of water and hydrocarbon liquids. The resultant frothover caused the sinking of the internal roof and overpressured the tank. The seam between the fixed roof and shell ruptured. The subsequent evolution of hydrocarbon vapour affected 19 contractors who were treated for nausea, headache and respiratory problems. It was estimated that 87 barrels (10 tons) of gas oil and slop oil vaporized to atmosphere or was released to the tank dike/bund area.

One way of identifying potential problem areas in the operation of slops tanks is to carry out a HAZOP study. In this way pressure, temperature, flow and composition variables can be considered in a logical and systematic manner.

When did you last perform a HAZOP study on your slops systems?

2.6 Usual sources of water in tankage

The most common source of water is from leaking steam coils or suction heaters. Water may also enter the tank through the roof hatch, holes in the roof, around the gauge cables, or through any other unprotected openings. Periodic checking and good maintenance of tanks and heaters will do much to eliminate this hazard.

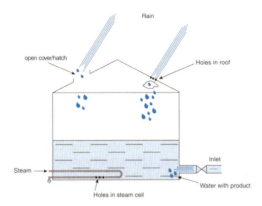

Checking for water in the tank and draining tanks are two important tasks for the following reasons:

- water can cause water explosions in process units;
- water can cause many problems for the customer, such as freezing in winter;
- the customer does not want water to be mixed with product.

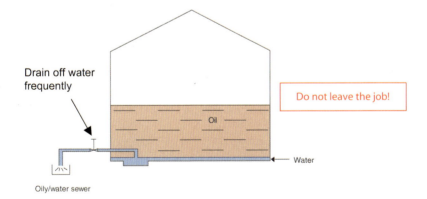

Never leave a tank unattended when draining off water.

Never leave a tank draining job unattended. You could lose the contents of the tank to the water treatment plant and overload the system.

Ensure draining systems that might freeze in cold weather are adequately winterized and the insulation and steam/electric trace heating are maintained in good condition. Excessively rapid water draining rates can result in product being drawn out into the sewer system. Reduce the rate of drainage to allow operators to check if significant quantities of water remain in the bottom of the tank.

When water is drawn from naphtha or gasoline tanks, product can easily be mistaken for water unless some special means of detection is used. For example, wood tends to absorb naphtha or gasoline, whereas a drop of water tends to stand on the surface. Therefore, frequent testing of the stream with a stick or small board prevents losing oil to the sewer. The use of detection paste is another method for checking water drawn from tanks. The paste changes colour in the presence of water.

Automatic water draw-off valves are available which close on the detection of the hydrocarbon/water interface. The integrity of these systems must be guaranteed before allowing the operator to leave the draining operation otherwise excessive hydrocarbon liquid could enter the water treatment system.

3

Sources of ignition

3.1 Static electricity

3.1.1 What is static electricity?

Static electricity is a surface phenomenon associated with the contact and separation of dissimilar insulating surfaces. It can occur at solid-solid, solid-liquid or liquid-liquid interfaces.

When two *dissimilar insulating materials* touch and separate, a charge imbalance occurs. The surface that tends to hold electrons more tightly will 'steal' charged particles from the other surface. The migration of charges results in the surfaces becoming oppositely charged. Static electricity can lead to a spark discharge to rectify the imbalance of charges.

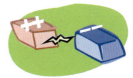

An example of an everyday static electricity occurrence:

A static spark may jump from the end of a finger to a metal switchplate after one has walked across a carpet.

3.1.2 Conductivity

Conductivity refers to the ability of a liquid to diffuse an electrostatic charge. Its inverse form is called resistivity, the ability to retain an electrostatic charge. Conductivity is measured in picosiemens per meter (pS/m) while resistivity is measured in ohm-cm.

Units
$1 \text{ pS/m} = 10^{-14} \text{ ohm}^{-1} \text{ cm}^{-1}$.
As resistivity, this is equal to 10^{14} ohm-cm

Substance	Conductivity
Hydrocarbon distillates (gasoline, kerosene, white spirits, jet fuels, napthas, diesel oils)	• Low conductivity • 0.1–10 pS/m. • Strong electrostatic accumulators
Residues and crude oils (black oils, asphalts)	• Higher conductivity than hydrocarbon distillates • 10^3–10^5 pS/m • Any charge generated is rapidly dissipated
Distilled water	• High conductivity • 10^8 pS/m

Generally (when all other conditions are similar), the lower the conductivity of a liquid, the more significant are the dangers involving static electricity.

However, sparking from a surface of a liquid is a significant risk irrespective of the liquid's conductivity if it is discharged into any insulated metal container. Hence the importance of earthing/grounding and bonding metal containers before commencing the filling operations (see section 3.1.6 on Static control on page 42).

Conductivities of some flammable chemicals

Liquid	Conductivity, σ (pS/m)	Liquid	Conductivity, σ (pS/m)
Acetone	5×10^8	Toluene	1
Methyl ethyl ketone	5×10^6	Xylene	0.1
Ethyl benzene	30	Heptane	3×10^{-2}
Styrene monomer	10	Benzene	5×10^{-3}
Cyclohexane	2	Hexane	1×10^{-5}

Note: $5 \times 10^8 = 500,000,000$
$1 \times 10^{-5} = 0.00001$

Electrostatic accumulation is significant *unless**:

• **Conductivity exceeds 50 pS/m (resistivity less than 2×10^{12} ohm cm); and**
• **Chemical is handled in earthed/grounded conductive containers.**

****Not applicable for mists**

3.1.3 Static electricity generation

Static charge in the petroleum industry results from contact and separation in flowing liquids. Liquid flow through a pipe is a common static generating situation.

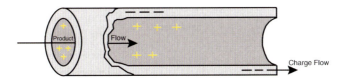

The presence of impurities, such as water, metal oxide and chemicals, increases the static generation characteristics of a liquid.

Static producing situations	Examples
Liquid flow through pipes and filters	Fuelling of vehicles, loading of tanks
Liquid agitation and mixing	Jet mixing
Free fall of liquids	Overhead filling of tanks
Splashing by the break-up of jets or bubbles	Splash filling of tanks, high speed ejection of liquids from nozzles
Gas bubbling through liquid	Air in a liquid rises to the surface
Settling of droplets of one liquid through another	Water droplets separating out in a tank containing a petroleum liquid
Settling of solid particles in liquids	Rust and sludge particles settling in a tank
Impingement of solids on solids	Sand blasting, impingement of solid particles on tank plates

3.1.4 Static build-up during transfer operations

The figure below shows the static build-up in the oil as it moves along the line from uncharged oil in one tank to charged oil upon reaching its destination. (Refer to **API 2003** for further details.)

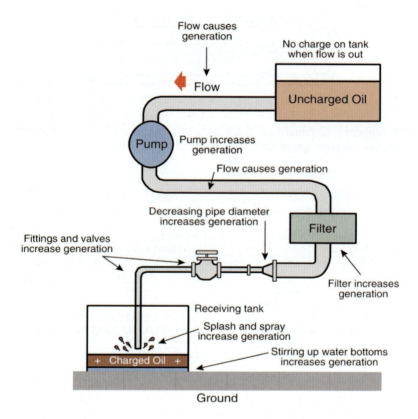

Static build-up on a line

Larger electrostatic charges are obtained with:

- filters;
- pumps;
- lower conductivity liquids;
- smaller diameter pipes;
- presence of water and other particulate matter;
- splash filling;
- increased velocity/flowrate.

3.1.5 Why is static electricity dangerous?

The principal hazard of static electricity is a spark discharge in a flammable atmosphere.

**Static electricity discharge + flammable atmosphere =
DANGER OF FIRE & EXPLOSION**

Static discharge has often been given as the cause, or likely cause of fire and explosions, when circumstances have excluded the possibility of more familiar forms of ignition. They have occurred during many different operations such as steaming out of equipment, tank filling operations and sampling using plastic containers.

In order for an electrostatic charge to become a source of ignition, four conditions must be fulfilled:

- **electrostatic charge generation;**
- **accumulation of an electrostatic charge capable of producing an incendive spark;**
- **a means of discharging the accumulated electrostatic charge in the form of an incendive spark, such as a spark gap;**
- **an ignitable vapour-air mixture in the spark gap.**

ACCIDENT Explosion triggered by a static electricity discharge!!!

Approximately five minutes after the start of transfer of reformate into a gasoline tank for a blend of unleaded motor gasoline, a noise (boom) was heard in the vicinity. The transfer pump was stopped immediately and roof distortion was observed. The mild explosion was thought to have been triggered by a static discharge. The high pumping rate caused the accumulation of static charges while *natural breathing* due to temperature changes produced a flammable atmosphere. The presence of a flammable atmosphere during a spark discharge can lead to fire and explosion incidents.

ACCIDENT A 80,000-barrel (13,000 m³) storage tank exploded and burned as it was being filled with diesel. The tank had previously contained gasoline, which had been removed earlier in the day. The tank contained approximately 7,500 barrels (1,200 m³) of diesel at the time of the explosion. The resulting fire burned for about 21 hours and damaged two other tanks. Nearby residents were evacuated, and schools were closed for two days.

The operations allowed the tank to be switch loaded (definition on page 51) at flow velocities significantly higher than those in both its own procedures and industry-recommended practices. The high velocity of the diesel in the tank fill piping and the turbulence created in the sump area resulted in the generation of increased static charge and, combined with the very low electrical conductivity (static accumulating) liquid, an elevated risk for a static discharge inside the tank.

Tank operations before the fire, which included a partial draining that landed the floating roof and partial fillings before draining dry, allowed a large amount of gasoline vapour to be generated and distributed within the tank to create a flammable fuel-air mixture both above and below the floating roof.

View of fire and what was left of the tank

All the conditions necessary for fuel vapour ignition were present in the storage tank at the time of the accident, and the explosion most likely occurred when a static discharge ignited a flammable fuel-air mixture in the space between the surface of the diesel and the floating roof. The extensive damage to the tank is consistent with the flammable fuel-air mixture above the floating roof contributing to the force of the explosion.

Lessons learned

Storage tank operating procedures should include instructions:

- for minimizing the possibility of creating a flammable atmosphere and the occurrence of a static discharge inside a tank after a floating roof has been either intentionally or unintentionally landed, especially for tanks where switch loading is likely to occur.

- to ensure that product flow rates in both the tank fill line and the discharge nozzles are restricted to provide a level of protection against excess static electricity.

continued

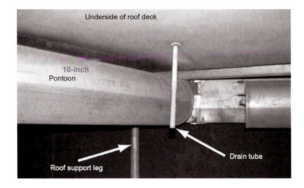

Drain tube extending down from floating roof tank—acts as a spark promoter

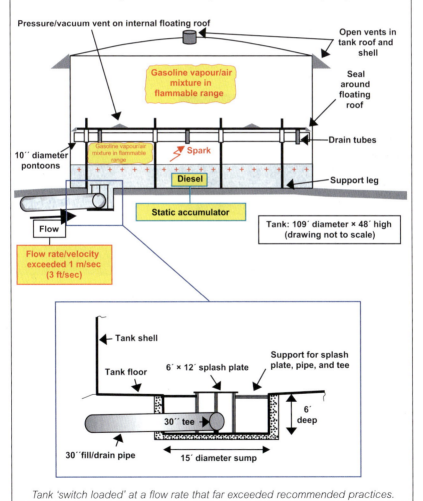

Tank 'switch loaded' at a flow rate that far exceeded recommended practices.

3.1.6 Static control

The options available to manage the risk of fire and explosion are:

1) Preventing static build-up:

    a)   earthing/grounding and bonding;

    b)   reducing static generation;

    c)   using anti-static additives.

AND/OR

2) Eliminating flammable atmospheres:

    a)   purging;

    b)   inerting;

    c)   eliminating the vapour space.

1)   Preventing static build-up

a)   Earthing/grounding and bonding

Bonding of floating roof

Tank grounding/earthing

Note: *The type of cables shown in the pictures above are designed to deal with static accumulation. They cannot cope with sudden huge electrical discharges such as lightning as demonstrated in the picture on the right which shows a floating roof to shell shunt test (submitted to a 830 A current to simulate lightning). Note the sparks generation—wax and rust deposits increase sparking.*

Picture from tests by Culham Electromagnetics and Lightning Limited for the Energy Institute (UK) and the API.

Tank not earthed/grounded and not bonded to fill-pipe.

When a charged stream enters a metal container or tank, it induces charge separation on the tank wall. A charge equal in magnitude to the fluid charge but of opposite sign will be induced on the inside surface of the tank. A charge of the same sign as the incoming stream will be left on the outside surface of the tank.

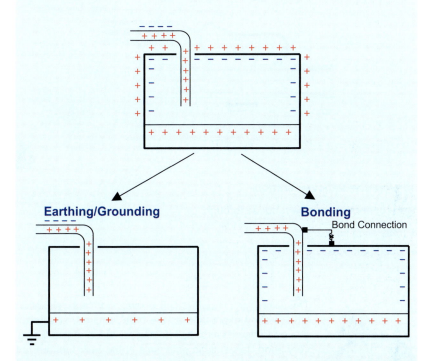

Earthing/Grounding

Bonding

Tank earthed/grounded but not bonded to fill-pipe.

When a tank is grounded, electrical charge 'drains off' from the conductor (tank) and flows to ground. The fill-pipe surface remains charged.

Tank not earthed/grounded but bonded to fill-pipe

When a tank is bonded, the charge is 'neutralized' between the tank and fill-pipe surfaces. The charge on the outside surface of the tank combines with the charge on the outside surface of the fill-pipe. The charge within the tank and the fill pipe remains.

Earth/ground and bond all equipment and containers to prevent static charge build-up.

Note that it is not normally necessary to install bonding cables across flanges because both sides of the pipe are at the same potential (earth/ground) and the liquid flowing through the pipe cannot form an insulating barrier.

However in the case of material that could polymerize and cause an insulating barrier, an electrical bond across the joint will be necessary.

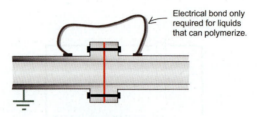

Electrical bond only required for liquids that can polymerize.

Hoses should be electrically continuous.

b) Reduce static generation

Static generation can be reduced through:

• relaxation time;
• good practice.

The discharge process is called *relaxation*, and relaxation time is often expressed as the time required for a given charge to decrease to half its original value. If this time is very short, large static potentials in the bulk fuel are not created because the relaxation process takes over and controls the charge that can build-up. Relaxation time depends primarily on the conductivity of the liquid. It can vary from a second to many minutes.

Residence time is the time to ensure that charge on the liquid is relaxed to a value which is considered safe after flowing through filters, pipes, valves, etc. into vessels and tanks. It can be longer than the measured relaxation time.

Residence time/loading rates to reduce charge generation into tanks

Condition	Measures
Free water is present in oil	• Restrict linear velocity in the pipeline to **1 m/s** (3 ft/s) (this applies for the complete transfer of wet product).
Product is pumped through a filter	• Allow a residence time of at least **30 seconds** for hydrocarbon oils of conductivity **2 pS/m** and higher • Provide for residence time of up to **100 seconds** for lower conductivity oils. • From a technical point, the relaxation time may be disregarded for products having conductivity greater than **50 pS/m**. This conductivity may be inherent in the liquid or achieved through the use of antistatic additives.
Conductivity has not been measured	• Allow a residence time of at least **100 seconds**.
Mists and splash filling	• Limit the velocity of the incoming liquid stream to **1 m/s** (3 ft/s) until the fill outlet is submerged or in the case of floating-roof tanks, until the roof is afloat.

Note: Relaxation chambers are often installed after filters in pipelines.

Measurements have shown that a filter can produce from 10 to 200 times more charge than a system without filtration.

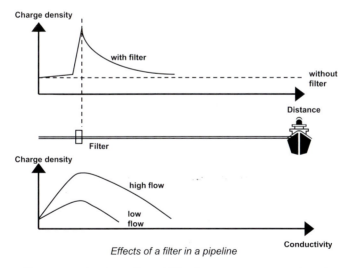

Effects of a filter in a pipeline

Filters with a pore size less than 150 microns may generate a hazardous charge level, particularly if they are partially plugged, while wire screens with pore size exceeding 300 microns (50 mesh per inch) are unlikely to generate hazardous charge levels.

It is poor practice to install fine filters at the discharge end of loading arms/hoses because there is no residence time to dissipate the accumulated charges in the liquid before entering the vessel/road truck.

Good practice

- Provide bonding between source of charge and tank.
- Provide earthing/grounding to the tank shell.
- Maintain multiple stainless steel shunts on floating roofs in good condition.
- Maintain all bonding connections on floating roof tanks in good condition.
- Avoid overshot filling (filling through top inlets with free fall of liquid).
- Ensure that the outlet of the fill pipe discharges at the bottom of the tank with minimum agitation of the water and sediment on the tank bottom.
- Do not use siphon breakers, which permit air or vapour to enter the 'downcomer', where the outlet of the fill line is attached to a downcomer.
- Avoid discharging product from a swing line elevated above the liquid level.
- Drain free water from the tankage regularly.
- Avoid pumping substantial amounts of air or other entrained gas into the tank through the liquid.
- Eliminate the presence of any objects that are not grounded such as loose gauge floats in the tank.
- Eliminate the presence of spark promoters.
- Take adequate protective measures during tank dipping or sampling. A detailed discussion may be found in *Sampling, dipping and taking temperatures* in Section 3.1.10 on page 54.
- Avoid landing the floating roof on its legs (becomes a fixed roof tank) except when it is taken out of service for maintenance.

c) Use anti-static additives

Static electricity build-up can also be prevented by the use of antistatic additives. Antistatic additives work by increasing the conductivity of a liquid.

2) Eliminate flammable atmospheres

The best approach to prevent ignition by static charge is through the elimination of flammable atmospheres. This can be achieved through:

a) purging

b) inerting

c) eliminating the vapour space

Purging and inerting can be performed using nitrogen gas or other inert gases. Refer to BP Process Safety Booklet *Hazards of Nitrogen and Catalyst Handling* for information on handling nitrogen safely.

Eliminate flammable atmospheres!

ACCIDENT **Flammable atmosphere results in foreman catching fire!!!**

An installation yard foreman was working in the electrical workshop when he decided to visit the drum filling area. At the drum filling area, he was informed that the weigh scale was not operating correctly. He went to have a closer look at the weigh scale and stepped on it, reaching at the same time with his hand to get hold of the fill pipe for support. Before his hand could reach the pipe there was a flash below his feet and flames engulfed the scale as spilled fuel in the pit below caught fire.

Investigations revealed that the foreman was wearing shoes with plastic soles. Due to the lack of any other possible source of ignition, it was concluded that the man's body accumulated a static charge whilst travelling to the shed. This electrical charge was discharged from his body to the scale, which was correctly bonded to the pipes and earthed/grounded.

Two important housekeeping lessons can be learned from this incident.

- Do not allow flammable products to accumulate around equipment, particularly in pits, due to the danger of fire should a source of ignition present itself.

- Maintain filling nozzles, loading arms etc. adequately to ensure that leaking valves and joints cannot contribute a further source of fuel in the event of a fire.

Humans are sometimes the source of ignition of flammable atmospheres. The minimum ignition energy of propane, n–butane and n–hexane is only 0.25 mJ while the energy released by human static discharge can reach 60 mJ.

0.25 mJ is equivalent in energy to dropping a dime (2.3 g) from a height of 1 inch (2.5 cm)!

The wearing of conductive or anti-static footwear can prevent the accumulation of an electrostatic charge on the person.

Footwear—conductive or anti-static?

One of the additional safeguards for personnel such as a tank dipper or sampler, who may cause or receive an accumulation of static electricity by various mechanisms, is the wearing of conductive or anti-static footwear.

Two degrees of conductivity are recognized—'anti-static' and 'electrically conductive'. Good electrically conducted footwear should have a resistance less than 1.5×10^5 ohms while anti-static footwear should have a resistance between 5×10^4 ohms and 5×10^7 ohms.

Anti-static footwear has sufficient conductivity to dissipate electrostatic charges and sufficient resistance to give protection against shock from electrical apparatus up to at least 250 volt. If anti-static footwear is used, regular testing with a special testing machine may be required to ensure that its conductivity remains adequate.

Refinery operator testing his shoes at the beginning of his shift.

Remember! You cannot see a static charge build-up. It is too late when a spark is produced. Observe your tank operating procedures and report all incidents of static electricity.

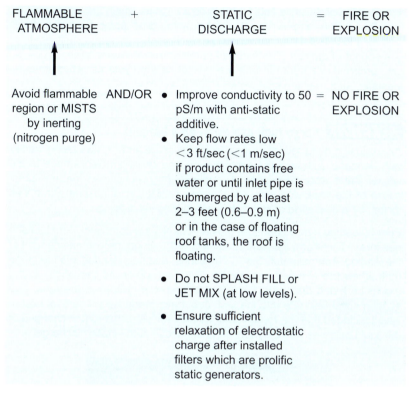

FLAMMABLE ATMOSPHERE	+	STATIC DISCHARGE	=	FIRE OR EXPLOSION

Avoid flammable region or MISTS by inerting (nitrogen purge)

AND/OR

- Improve conductivity to 50 pS/m with anti-static additive.
- Keep flow rates low <3 ft/sec (<1 m/sec) if product contains free water or until inlet pipe is submerged by at least 2–3 feet (0.6–0.9 m) or in the case of floating roof tanks, the roof is floating.
- Do not SPLASH FILL or JET MIX (at low levels).
- Ensure sufficient relaxation of electrostatic charge after installed filters which are prolific static generators.

= NO FIRE OR EXPLOSION

ACCIDENT Seal oil tank explosion due to splash filling!!!

An ignition of a hydrocarbon gas/seal oil mist air mixture by a static discharge resulted in an explosion within a seal oil tank.

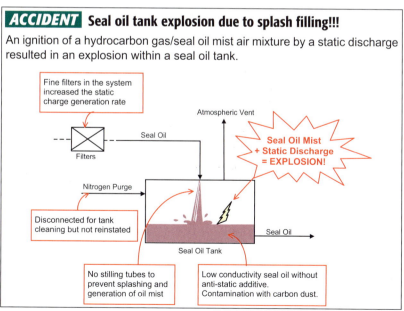

Fine filters in the system increased the static charge generation rate

Atmospheric Vent

Seal Oil

Filters

Seal Oil Mist + Static Discharge = EXPLOSION!

Nitrogen Purge

Disconnected for tank cleaning but not reinstated

Seal Oil

Seal Oil Tank

No stilling tubes to prevent splashing and generation of oil mist

Low conductivity seal oil without anti-static additive. Contamination with carbon dust.

Non-volatile combustible liquids can produce mists, which are easily ignitable even though the temperature is well below the flash point. Fine mist droplet build-up and splash filling can create an electrostatic discharge sufficient to act as a source of ignition. The nitrogen purge and inlet/return stilling tubes in a gas compressor's seal oil tank are critical safety systems, which must be reinstated after maintenance work.

ACCIDENT Cargo tank explosion on a barge!!!

An explosion occurred in a cargo tank of a barge while it was being loaded with benzene. Although the explosion caused considerable damage to the vessel, there were no injuries. The investigation concluded that the direct cause of the explosion was a flammable aerosol/air atmosphere that was ignited by a discharge of static electricity during initial filling.

Prior to loading a new chemical, the cargo tanks are completely emptied and gas freed. Therefore, the tank contained 100% air on the commencement of the loading. Approximately 3m³ (790 gallons) had been loaded into the cargo tank when the explosion occurred. Evidence points to the origin of the explosion being less than 1.4m (4.6 ft) from the bottom of the tank. The sudden initial flow rate into the tank may have produced an aerosol mist. It was concluded that the formation of an aerosol with subsequent charging of the liquid particles was the most probable cause of the static discharge.

Damaged cargo tank

To prevent the production of a mist (splash filling) inside a tank or a static discharge, it is good practice to restrict the filling velocity in the pipeline to 1 m/s (3 ft/s) until the outlet of the fill line has been covered by a minimum depth of 0.5m (1.6 ft). It is better practice to avoid going into the flammable region of vapour/air mixture by first inerting the tank to remove the oxygen content to below 5%.

Note: Benzene has a flash point of −11°C (12°F), a flammable range of 1.4 to 8.0% and low conductivity of 5×10^{-3} pS/m. It is therefore a static accumulator requiring special precautionary measures to avoid a potential ignition risk.

3.1.7 Switch loading

Switch loading refers to the loading of a low volatile product into a compartment or tank that previously held a high volatile product. An example of switch loading, which has been recorded in numerous incidents, is the loading of diesel (high static generator) into a compartment that previously held gasoline (presence of a flammable atmosphere).

The loading of a low conductivity combustible oil into a compartment or tank that has previously contained a flammable liquid provides the ideal conditions for an explosion. It is therefore important to remove the flammable atmosphere prior to loading the low conductivity liquid.

An estimated 70% of road truck incidents involving static are attributed to switch loading. Special precautions must be taken when switch loading is proposed.

ACCIDENT **Fatality during switch loading of semitrailer!!!**

Whilst top loading diesel fuel, a major explosion and fire erupted, completely destroying a semitrailer and killing its driver instantaneously. The driver had just commenced loading the first compartment, which had contained gasoline on the previous delivery.

On investigation it was found that:

- There was no evidence of compartment flushing;
- It could not be established whether the vehicle had been externally earthed/grounded;
- The loading rate was equivalent to 4.9 m/s (16 ft/s);
- Vehicle compartment (7.6 m^3 (2000 gallon) capacity) had been loaded to approximately 1.1 m^3 (300 gallons);
- Ambient temperature was 0.6°C (33°F) (low humidity);
- It was not conclusively established whether the filling downpipe was completely inserted into the compartment.

The incident was associated with a high filling rate at low ambient temperatures, providing conditions for a high static charge. It seems likely that a spark jumped across the point where the fill pipe did not approach the bottom of the compartment.

The company concerned made the following recommendations:

- inject anti-static additives to all distillates before loading;
- vapour-test vehicle compartments before loading distillates;
- reduce loading rates to the equivalent of 3.7 m/s (12 ft/s);
- increase loading supervision;
- install anti-static hydraulic valves on each filling downpipe used on distillate service.

53

The hazards of switch loading for trucks can be controlled/removed through:
- **relaxation time;**
- **good practices;**
- **vapour-space testing prior to switch loading to assess for the presence of a flammable atmosphere;**
- **removal of flammable atmosphere by ventilation or inerting;**
- **purge and gas-free the vessel prior to switch loading.**

Switch loading is a potentially hazardous operation.

3.1.8 Agitation

Air-blown batch agitators generate static electricity prolifically. Charges increase immediately after agitation has stopped and after the air mixing has been turned off. Initially, the charges can be quite intense, resulting in sparks darting across the surface of the oil. They gradually diminish over a period of 5–10 minutes.

Methods of protection include:

- inert gas blanketing;
- a closed circulating system that eliminates the free oil surface.

Poor practice: air agitation

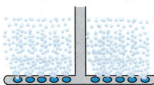

3.1.9 Mixing

Jet mixing

There is little evidence of static generation when using conventional low speed propeller mixing. However, jet mixing operations and high speed propeller mixing can cause a build up of static electricity through splashing and turbulent flow. Static generation is further increased if the product has a low conductivity, contains free water or when water bottoms are disturbed. Ignition may be possible if flammable mixtures exist at the surface.

Methods of protection:

- do not jet mix if vapour space contains flammable gas/air mixture;
- do not allow jets to break the liquid surface;
- perform mixing operations in floating roof tanks;
- otherwise, use inert-gas blanketing.

Agitation and mixing

ACCIDENT Tank explosion during cleaning!!!

An MTBE tank explosion occurred during a tank cleaning job. When the cleaning started, the vapour in the tank contained 73% nitrogen and 27% MTBE. During the last phase of cleaning, the vacuum truck started to suck vapours from the tank. This caused some underpressure in the tank. Air entered the tank through the top manhole forming an explosive mixture. The explosive limits for MTBE in air ranges from 1.5% to 8.5%.

When most of the liquid was removed, a high pressure cleaning device (rotating high pressure water nozzle) was put into the top manhole. Shortly after, an explosion in the tank occurred. The operator was blown off the roof and was most likely killed by falling from a height of 10m. The tank roof (1.4 metric tons [3100 lb]) landed 100m (330 ft) away between two other tanks.

The most likely source of ignition was a spark caused by a static discharge from the high speed fine particle water mist created by the high pressure head.

ACCIDENT The roof of this tank was damaged during an earthquake. Foam was applied as a preventive measure using foam pourers but the foam blanket was not maintained. Ignition occurred because of static build-up where a foam pourer maintained a continuous dripping of water and foam onto the naphtha surface.

Refer to BP Process Safety Booklets *Hazards of Water* and *Liquid Hydrocarbons Tank Fires: Prevention and Response* for more details.

3.1.10 Sampling, dipping and taking temperatures

A spark may be discharged within a flammable atmosphere during dipping or sampling. Metallic or conductive hand gauging tapes, sample cans or bottles on chains can act as spark promoters.

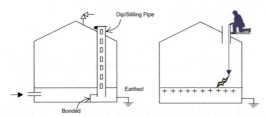

The tank diagram on the left represents the safer design of the two tanks. It is earthed/grounded and bonded and provides a method to dip the liquid where there is a shorter path for the discharge of any static build-up. Wait the stipulated time before sampling/dipping to ensure that any charge accumulated has dissipated to a safe level.

Example of an electronic temperature probe with bonding cable to be attached to the tank nozzle/hatch.

Note: Ensure good bonding to tank when carrying out these operations but remember that a safe period must elapse to ensure that accumulated charges on the surface of the liquid have dissipated. Ensure dip hatches are kept tightly closed when not in use in case of lightning strikes.

ACCIDENT **Fire caused by ineffectively earthed/grounded metal bucket!!!**

A fire occurred at a process plant when the discharge from a drain valve was collected in a metal bucket suspended from the drain valve body. The bucket was not effectively earthed/grounded since there was a thermoplastic sleeve around its handle preventing metal-to-metal contact. The assembly was also some 3m (10 ft) above ground level.

Oil had sprayed from the drain valve (open about 20%) for less than a minute when ignition took place. The flames reached about 3m (10 ft) high. The fire was confined to the drain bucket, quickly extinguished and the plant shut down. There was no re-ignition, and the fire caused only superficial damage.

The most likely cause for ignition was a static discharge, arcing from the metal bucket to the drain nozzle. The resulting spark being of sufficient energy to ignite the gaseous mist produced by the high pressure spray from the drain valve. The other possibility considered was a static charge on the operator who opened the drain valve.

Unacceptable practice: Draining into open buckets

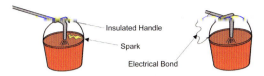

Insulated Handle
Spark
Electrical Bond

The collection of drainings from pipeworks/loading arms must be properly designed to include earthing/grounding and bonding and the correct container for the duty.

Precautionary measures during sampling and dipping

- Avoid sampling or dipping during loading or immediately after loading (especially low conductivity products not dosed with anti-static additive). Wait at least *30 minutes* for surface charge to dissipate.
  The decay of field strength is slower in larger tanks than in smaller ones. The slow decay is probably due to further charge generation when small charged particles of water, dirt or other materials settle in a large tank.

- No waiting period is required theoretically when using completely nonconductive hand gauging or sampling devices. However, it may not continue to retain the necessary high degree of insulation when exposed to environmental situations. So always wait the stipulated time to be on the safe side.

- Automatic gauging devices can be used safely in vessels containing static-accumulating oils with flammable atmospheres. Floats should be electrically bonded to the vessel shell through lead-in tape and/or guide wires.

- Manual dipping of tank while receiving should only be permitted if the liquid is of high conductivity (above **50 pS/m**).

- Avoid dipping or sampling tanks with flammable atmospheres or with mist above the oil level.

- Use only natural fibre cord for sampling or dipping. Synthetic (nylon and polyethylene) ropes have been shown to charge an insulated person when it slips rapidly through gloved hands for appreciable distances.

- Use non-sparking sample containers and dip tape plummets (for example, brass).

- Use only metal containers that are electrically bonded before taking samples.

- Make sure the steel dip tape is in contact with the metal edge of the dip hole.

- Earth/ground sampling tapes before introducing to tank headspace.

- Use only approved electrical equipment including lamps, torches and radios.

- Do not dip or sample during electrical storms, heavy rain or hail.

- Wear proper Personal Protection Equipment such as non-slip and anti-static footwear, goggles and gloves.

- Close dip hatches properly after completing the task.

3.1.11 Steaming

Steam is frequently used for cleaning equipment and plant of flammable material before it is certified safe to be taken into workshops for hot work or sent for scrap. Some precautions need to be taken to avoid static hazards during steaming.

Precautionary measures that can be taken include:

- earth/ground the vessel/equipment;
- bond the steam hose nozzle to the vessel electrically;
- earth/ground the steam pipe and ensure that the flexible hose is electrically conductive;
- emit steam very slowly at the beginning of the steaming out operation;
- take adequate precautions to avoid blowing wet steam into the vessel;
- wear anti-static or conducting footwear.

Refer to BP Process Safety Booklet *Hazards of Steam* for additional information on the subject.

3.2 Other sources of ignition

Only in approximately 50% of reported fires and explosions is the source of ignition identified with any degree of certainty. This leads to the view that when there is a release of flammable material it is likely to find a source of ignition. It is essential therefore to strictly control the use of any potential sources of ignition on any premises holding flammable materials.

Two sources of ignition have been discussed in detail in other sections of this booklet, Section 3.1 covers *Static Electricity* (page 35) and Section 5.8 covers *Stray Currents* (page 90).

Numerous other potential sources of ignition exist. They include:

- smoking and flames;
- lightning;
- internal combustion engines;
- hot work such as welding and grit blasting;
- maintenance of any electrical equipment including cathodic protection in classified hazardous areas;
- use of any electrical equipment not meeting the specification for the classified hazardous area (see the next example) including portable lamps;
- frictional ignition such as that associated with the impact of light alloys of aluminium and magnesium on rusty steel known as the 'thermite' reaction;
- pyrophoric material;
- boilers and furnaces;
- hot pipes/vessels.

To read more on ignition sources and on fuel characteristics, refer to *Sources of Ignition* by John Bond, ISBN 0 75061 180 4.

ACCIDENT A 5000 barrel crude tank was being cleaned when an explosion lifted it several feet off the ground, splitting the roof open 1/3 to 1/2 the circumference at the roof seam and shooting a yellow flame horizontally 20 to 30 feet out of the roof opening. The vapours coming from an open hatch ignited on the 300 v DC line that was left seven years before when an ultrasonic level sensor was dismantled.

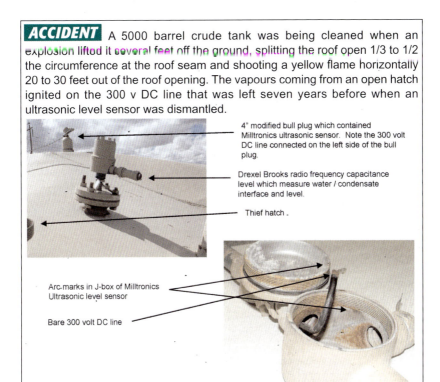

4" modified bull plug which contained Milltronics ultrasonic sensor. Note the 300 volt DC line connected on the left side of the bull plug.

Drexel Brooks radio frequency capacitance level which measure water / condensate interface and level.

Thief hatch .

Arc-marks in J-box of Milltronics Ultrasonic level sensor

Bare 300 volt DC line

ACCIDENT A contract welder was performing welding/grinding during the installation of a safety gate on a recently constructed catwalk at the top of a produced water disposal tank. The wind blew the sparks onto the adjacent tank and into a pressure compensating valve. Ignition of the tank atmosphere caused an explosion resulting in significant damage to the roof and shell.

3.2.1 Restricted areas

It is good practice for each facility to define the physical limits where the use of sources of ignition are restricted or prohibited. This area normally extends off-shore around jetties at terminals.

A Hot Work Permit is issued to authorize specific activities such as:

- use of non-approved lights or non-intrinsically safe electrical equipment;
- work on flameproof or explosive-proof equipment/enclosures and Cathodic Protection Systems;
- any maintenance activity involving or producing a source of ignition, such as welding, grit blasting, grinding, pneumatic drilling, etc.

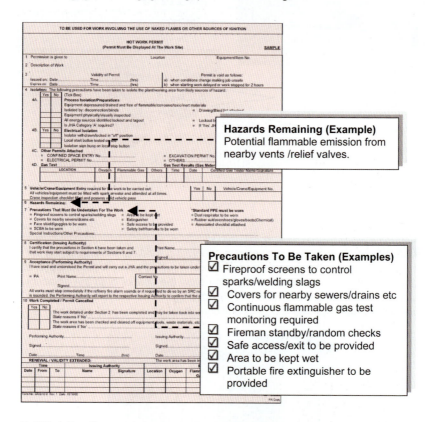

Hot Work Permit

In addition, personnel driving trucks carrying flammable liquids should not be carrying cigarette lighters and matches.

> **ACCIDENT** **Dropped cigarette lighter causes service station fire!!!**
> A tanker driver was making a night delivery of gasoline to a service station. As
> he opened the man-pit cover, there was an explosion, followed by a flash
> fire. The fire was quickly put out, but the driver sustained slight burns to his
> face. Later his cigarette lighter was found in the man-pit. It seems that it
> dropped from his pocket and sparked on hitting the ground.

3.2.2 Use of diesel driven equipment

Each facility must have a policy and procedures covering the use of diesel
engines. Unless the equipment meets the area classification where it is likely to
operate, its movement must be strictly controlled, for example, through barrier
control and/or a Hot Work Permit. It is good practice to always locate
unprotected diesel engines outside a classified hazardous area when
undertaking maintenance work. Refer to the Classified Hazardous Area
Drawing.

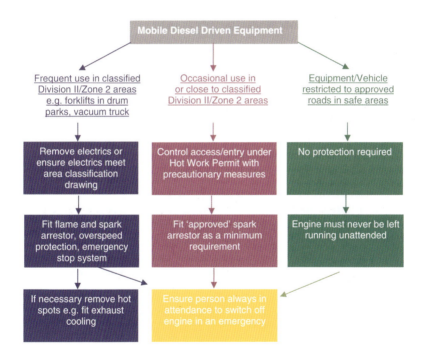

ACCIDENT **Car in vicinity of ship loading gasoline!!!**

The main hazard from petrol/gas driven vehicles is the spark engine system, dynamo and battery. These should be considered as 'sparking in normal use' and restricted to roads located in safe areas. It is not good practice to authorize the use of a gasoline driven engine in a classified hazardous area.

Petrol/gasoline driven car

Car access should not be allowed during loading/unloading of products at jetties

3.2.3 Vacuum trucks

Vacuum trucks are often used to remove oil/sludge/wastes from pits, tanks, sumps and gullies. There are inherent hazards associated with the use of these vehicles particularly if low flash point hydrocarbons could be transported or if the truck is expected to operate in a Division II/Zone 2 Classified Hazardous Area. It is important that:

- A flammable/air mixture or mist is not generated inside the vacuum tank which could be ignited by a number of ignition sources such as pyrophorics, stones or static.

- An unprotected diesel engine and ancillary components (electrics and hot surfaces) do not come into contact with a flammable atmosphere.

In general:

- Vacuum trucks must not be used to handle low flash point wastes unless specifically designed for such purposes.

- Diesel engines require to be fitted with a number of special devices/equipment such as over-speed protection if they are expected to operate regularly in a Classified Division II/Zone 2 Hazardous Area or could come into contact with a flammable atmosphere.

- Vacuum truck drivers/contractors must understand the properties of the waste material being transported and trained/competent in the procedures for ensuring safe loading/discharging of the tank contents.

ACCIDENT **Vacuum truck explosion with multiple fatalities!!!**

Three people were killed and four seriously burned when two vacuum trucks were unloading waste liquids to an open disposal pad. The waste contained low flash point gas condensate which formed a flammable vapour cloud. The cloud was ignited by one of the trucks' diesel driven engines which were left running.

Burned out vacuum trucks

ACCIDENT An explosion occurred within the engine compartment of a refinery pick-up truck. The pick-up truck was driven into and parked in a car park next to an operational vacuum tanker. At the time of the incident, pipework modifications associated with the gasoline storage facilities were being undertaken. The tanker was removing gasoline from an opened pipeline within the bund area of a gasoline storage tank adjacent to the car park. As the pick-up truck stopped, the engine began to 'race' and seconds after the driver left the cab an explosion occurred within the engine compartment quickly followed by second flash within the cab of the vehicle.

One of the construction team working on the gasoline line extinguished the fire, using a portable dry powder fire extinguisher.

The pick-up truck was driven and parked within an unmarked and inadequately controlled hazardous area created by the venting of gasoline vapours from the operational vacuum tanker. The vapours entered the engine intake causing the engine to 'race'. These vapours ignited within the engine compartment causing an explosion.

Lessons learned

- Everybody involved in vacuum tanker operations should thoroughly understand the risks.
- Risk assessments should be accurate and specific for the task.
- The Permit to Work should be specific to the work in hand.
- Temporary hazardous areas resulting from maintenance work must be properly controlled.

3.2.4 Mobile phones
Use of mobile telephones in classified hazardous areas

It is recommended that mobile telephones and radio pagers are switched off in classified hazardous/restricted areas. However, there is no evidence, as yet, that mobile phones can ignite a flammable gasoline atmosphere under conditions normally found at retail outlets. It has not been proven that mobile phones will not ignite other flammable atmospheres like H_2/LPG.

Studies conducted on ignition incidents at fuelling/gas stations have found that static electricity, rather than mobile phones, is often the cause of many of the incidents that involve the presence or usage of these devices. The use of automatic latching on fuel nozzles is found to encourage drivers to sit in their vehicles during filling. On completion they return to remove the filling nozzle having accumulated a static charge, which flashes to earth at the filling point.

ACCIDENT **Fire at a retail outlet from static discharge!!!**

A teenage girl got out of her car at a retail outlet, started pumping gasoline using an automatic latching on the fuel nozzle, and got back into the car. When the lock clicked off, she got out of the car (having generated a charge by friction between her clothing and the car seat) without grounding herself. When she reached for the pump nozzle, a blue flame jumped from her hand to the nozzle. The car burst into flames. Fortunately, she was not injured.

Ignition caused by static discharge is most applicable in dry (low humidity) environments/climates

---

> **Do not use mobile phones where they are prohibited!**
> **If they are allowed, avoid being distracted and focus on the task at hand!**

Refer to the BP Process Safety Booklet *Hazards of Electricity* for additional information on static electricity and mobile phones.

4

Loading and unloading of road and rail tankers

4.1 Loading and unloading

Road and rail tankers are generally used to transport product from refineries to marketing terminals or to customers in relatively smaller quantities (less than 100 tons). Tankers are filled at loading gantries either through top loading or bottom loading facilities with/without vapour recovery systems. The information in this chapter is applicable to road or rail trucks unless otherwise stated.

Top loading of a road tanker using a loading arm

Bottom loading of a road truck. End of hose fitted with a dry break coupling

Top Loading	<ul><li>An articulating arm is inserted into a tank compartment through a hatch on top of the truck.</li><li>A long fill pipe extends to the bottom of the compartment.</li><li>Liquid level rapidly covers the bottom opening of the fill pipe resulting in low vapour generation.</li><li>Reduced splashing minimizes generation of static charges.</li><li>Loader is more exposed to vapours.</li></ul>

Bottom Loading	<ul><li>A hose or flexible arm is attached to the bottom of the road tanker.</li><li>Vapour generation is minimized by the introduction of liquid through the bottom of the road tanker compartment.</li><li>Bottom loading has the disadvantage of slow electrostatic charge dissipation unless a conducting rod from the top to the bottom of the compartment is provided.</li><li>Requires a reliable level and shut-off system.</li><li>There are fewer opportunities for things to go wrong (than with top loading); for example:<ul><li>splash filling;</li><li>incorrectly positioned fill pipes;</li><li>introduction of debris into the top hatchway.</li></ul></li><li>No requirement to have a driver/operator standing on the top of the truck</li></ul>

Tankers can be loaded with closed or open to atmosphere configurations:

- Open configurations refer to those chemicals or petroleum compounds with vapour pressures <1.5 psia (0.1 bara). However, open top loading may not be permissible due to restrictions on emissions of Volatile Organic Compounds (VOCs).

- Closed configuration involves total vapour recovery (vapour balancing/return) (see *Section 9.5: Vapour balancing/return* on page 156).

Loading and vapour recovery arrangement for a railcar with integral level sensors

Whenever possible, bottom loading should be preferred when designing a new facility as it is the only foolproof engineering solution to falls from height during tanker loading. It also often provides quicker and safer escape routes than top loading.

ACCIDENT **Fatal fall from top of a truck!!!**

A driver died as a result of a fall from a fixed ladder on a tanker truck. The driver fell after he climbed the side ladder of the trailer to close the dome cover in preparation for departure.

4.1.1 Top loading

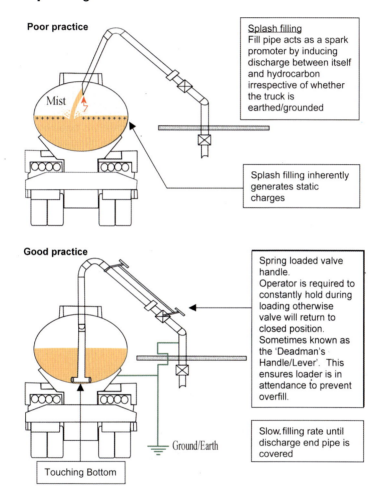

Poor practice

Mist

Splash filling
Fill pipe acts as a spark promoter by inducing discharge between itself and hydrocarbon irrespective of whether the truck is earthed/grounded

Splash filling inherently generates static charges

Good practice

Spring loaded valve handle.
Operator is required to constantly hold during loading otherwise valve will return to closed position. Sometimes known as the 'Deadman's Handle/Lever'. This ensures loader is in attendance to prevent overfill.

Slow filling rate until discharge end pipe is covered

Ground/Earth

Touching Bottom

Earth/ground proving systems are available to ensure that only when a good earth/ground is made are the loading pumps allowed to operate to fill the tanker.

Ensure the end pipe configuration (top loading) touches the bottom of the compartment to avoid splash filling

ACCIDENT **Road truck loading causes explosion and fatality!!!**

The driver of a road tanker was top loading diesel fuel into his tanker. He inserted a hose into the tank compartment that did not reach the tank bottom. The end of the flexible hose was approximately 2'6" (0.76 m) above the bottom of the tank compartment. The driver was standing on top of his vehicle watching the filling when an explosion and fire occurred. His clothing caught fire but he managed to climb down from the tanker. Fellow workers smothered the flames. However, he died from his burns two days later. The explosion was caused by an electrostatic charge igniting flammable vapours in the road tanker as it was being filled. *The vehicle had not been earthed/grounded. No earthing/grounding facilities were provided at the filling point to allow the road tanker to be bonded to earth/ground and the loading system was not designed to avoid splash filling.*

The precautions recommended include avoiding 'splash filling', and earthing during filling operations. Had even one of these precautions been taken, it is unlikely that the fire would have occurred, providing the filling rate was also controlled to avoid excessive generation of static electricity.

4.1.2 Bottom loading

Loading velocities for bottom loading are reduced because the static charge on the liquid product takes a longer time to dissipate as it needs to travel longer distances to the earthed/grounded sides of the tank compartment in the absence of a fill pipe. In top loading, the earthed/grounded metal fill pipe provides an immediate path for the dissipation of static charges.

> **In the case of bottom loading, unless there is an earthed conducting rod from the top to the bottom of the compartment, giving an electrical effect similar to the fill pipe, there is an increased possibility of electrostatic discharge and so the loading velocities are reduced.**

There are three major concerns when loading road trucks and railcars. These are:

- static electricity;
- overfilling;
- drive-aways.

4.2 Static electricity

> **THE FIRST TASK:** Earth/ground and bond tanker/road truck, equipment and containers to prevent static charge build-up.

Static electricity is a very important consideration when loading a road truck/car. Road trucks/cars must first be connected to earth/ground and bonded connections need to be made appropriately. This protects the vehicle from sparks, due to static charge accumulation, that can serve as an ignition source (see *Section 3.1: Static electricity* on page 35 for a detailed look into the issues related to static electricity).

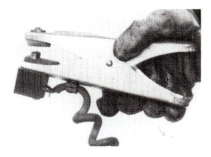

Earthing/grounding

Earth/ground proving system

The earth/ground proving system is self-checking and it permits product loading to/from truck only when satisfactory earthing/grounding is achieved. However, it does not indicate that the product loading hose and vapour recovery hose are electrically continuous.

Rail tankers have adequately low resistance to ground through the rail so as to prevent the accumulation of electrostatic charge of sufficient voltage to cause an incendiary spark. It is therefore unnecessary to bond the rail tanker or the rails to the fill pipe. However, there is a possibility of stray currents that can be controlled by bonding loading lines to the rails. This is preferable to bonding the loading lines to the rail tanker to assure a permanent bond and to avoid human error.

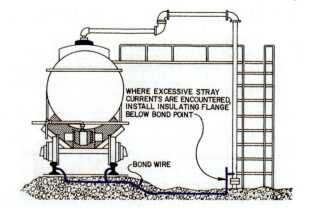

*Bond rail of tank
car loading spur
to piping*

WHERE EXCESSIVE STRAY
CURRENTS ARE ENCOUNTERED,
INSTALL INSULATING FLANGE
BELOW BOND POINT

BOND WIRE

Loading activities that can cause a static electricity hazard:

- splash filling;
- switch loading;
- sampling, dipping and taking temperatures.

Measures to be taken during loading of road and rail tankers
1) Ensure that equipment is grounded before commencing loading.

2) Do not exceed recommended flow rates.*
    • When top loading, or when bottom loading with a compartment with a central conductor (dip tube guide or dummy fill arm), limit Vd values to **0.5 m²/s** for road trucks (0.8 m²/s for rail cars).
    • When bottom loading without a central conductor, limit Vd values to **0.38 m²/s** for road trucks (0.6 m²/s for rail cars).
V = loading fill pipe linear velocity (m/s), d = the fill pipe diameter (m). These values are to be subjected to a maximum filling velocity of 7 m/s.

3) Only open the hatch of the compartment being loaded to reduce the amount of hydrocarbons emitted.
4) Avoid dropping items into the compartment. Do not load if the visual check on a compartment before inserting the fill-arm reveals that there is debris present.
5) Ensure that the fill-arm is centrally located and extends to the bottom of the compartment.
6) Allow sufficient time for charge dissipation before removing the loading arm.
7) Reservations regarding feed line filters must also be considered—the finer the filter, the higher the charge generated. Fine filters are defined as having a pore size below about 150 µm (micron), although much coarser filters can cause problems if they are partially blocked.

* Note: Some authorities/companies may wish to fill at different rates depending upon the conductivity of the product.

ACCIDENT **Explosion and fire occurs when loading a road tanker!!!**

A loader was killed when an explosion occurred as he was lowering a sample bottle into the truck's compartment during the loading of toluene. Despite having passed a written safety examination a month before, the loader carried out the loading operations in a very unsafe manner:

- He left his pick-up truck engine running at the site of loading.
- He failed to chock the wheels of the road truck.
- He failed to hook up the earthing connection to the tanker.
- He failed to hook up a safety line to his belt when he was on top of the truck.
- He had tampered with the spring loaded valve (deadman's handle) so that product would continue to flow when he went to get the sample bottle in its metal basket and then proceeded to lower it into the tank whilst toluene was still loading.

It is highly probable that the ignition occurred as a result of a discharge of static electricity from the surface of the liquid to the sample can. The driver of the vehicle ignored basic safety rules including the earthing connection and escaped with minor burns.

Toluene has a low conductivity (1 pS/m), a flash point of 48C (39°F) and UFL equilibrium temperature of 37°C (99°F), hence the atmosphere inside the tanker's compartment would normally be within the flammable range during loading. Some refinery procedures require a relaxation time of 10 minutes for toluene before sampling in road or rail cars.

Similar forms of loading malpractice have been reported. Direct or remote supervision, for example, by TV cameras can minimize these bad practices. A technical solution should be aimed for so that excessive loading rates, loading without being correctly earthed etc. become impossible in spite of the driver and loader. Purging the truck of air with nitrogen prior to loading these particular low conductivity products should also be considered as an option.

Use of nitrogen for inerting

The use of nitrogen to inert road/rail trucks to prevent a flammable atmosphere introduces the risk of asphyxiation. Refer to BP Process Safety Booklet *Hazards of Nitrogen and Catalyst Handling* for additional information on safety issues associated with the use of nitrogen.

ACCIDENT A repair yard worker was found dead inside a railcar sent to the yard for maintenance. The railcar had been purged and left full of nitrogen without adequate warning/communication of the railcar's status.

4.3 Overfilling

Level instrument sensors are located within tank compartments, which signal when the liquid level has reached the fill line. When a level sensor is actuated, the flow may be stopped through the metered valve or a separate valve, or the electric system may shut down, which in turn trips the control and/or the feed pump. An electrical plug is provided on each tanker to connect the level instruments directly to the liquid fill control system. Prior to loading, the terminal plug connection on the tanker is connected to the rack control system and the road tanker is grounded.

To prevent overfilling:

- install *high level alarms*;
- connect the *high level sensors* to an automatic shutoff valve;
- use *preset counting devices* that measure the volume of liquid.

- **Loading bays should be equipped with alternative escape routes with emergency stop buttons and an additional button located approximately 30 m (100 ft) away from the rack.**
- **Isolation valves should be located where they can be easily and safely reached under fire conditions or preferably each product line should be equipped with an emergency isolation valve that can be operated by a single switch from a remote location.**
- **The emergency system should be designed to 'fail safe', i.e. valves close in the event of air or power failure.**

Emergency shutdown button and emergency shower located in a readily accessible location with good signage.

ACCIDENT **Fatal terminal fire caused by failure of level sensor!!!**

The road tanker compartment already contained 1,300 gallons (4,900 litres) of gasoline when it entered the loading rack. The driver set the meter thinking the tank was empty. Due to failure of the liquid level sensor, the compartment was overfilled. A second vehicle, entering the loading rack, ignited the gasoline vapours. The whole facility became engulfed in flames. The fire killed one person and injured eight others. This incident demonstrates how quickly action has to be taken to stop other vehicles entering the area when a spillage occurs.

In the event of a loading rack fire or other emergency, such as a spillage, it is very important that all product flow to the facility is stopped and isolated immediately.

4.4 Drive-aways

Moving a vehicle while it is loading product or driving away with the hose still connected can have serious consequences. The breakage of hoses can result in spills that could ignite, endangering lives and damaging property. Even if a spill does not occur, the loading equipment may be badly damaged.

Therefore, the following immobility protection measures may be implemented:

- provide physical barriers;
- place chocks beneath the road or railcar wheels (see photograph below);
- interlock road truck/tank truck electrical system with the earth proving unit to ensure that the engine cannot start unless loading is completed and the attachments are removed.

Dry break couplings that disconnect the hose from the tanker if it moves away may be used as additional protection.

Chocks located beneath wheels of rail truck to prevent movement

Dry break coupling

ACCIDENT **Road tanker drives away with hose still connected!!!**

When loading was completed, the driver of a road tanker left his vehicle while waiting for the hose to depressurize. He forgot to disconnect the hose from his vehicle when he returned and drove away with the hose still attached. This caused considerable damage to the line although minimal product was spilled. Product availability was impaired for 24 hours and as a result, 100 tonnes shipment of product planned for dispatch was cancelled.

The incident occurred as a result of human error, inadequate operating procedures and equipment failure. The operator did not inspect the vehicle before the driver pulled away. Also, the system in place that should have made it impossible for the vehicle to drive away before the hose was disconnected had failed.

Where bottom loading of petroleum liquid tankers via hoses or connections for vapour-return systems is used, a higher level of protection is justified to ensure that 'drive aways' do not occur.

ACCIDENT **Road tanker fire caused by misunderstanding at loading rack!!!**

An incident occurred due to a misunderstanding between staff at a loading gantry and a road tanker driver. The latter moved his truck while it was still being filled with motor spirit. Although the pump was shut down almost at once, the spilled spirit ignited. The loading facilities were severely damaged, the road tanker destroyed, and the driver received severe burns.

Unprotected diesel engines are a source of ignition so there is a need to control their movement into/out of loading bays. In order to eliminate some of the hazard in this regard when loading is taking place a barrier, which could be in the form of a painted line on the ground, should be established at a sufficient distance from the loading bays at which road trucks with engines stopped should be required to wait until space is available in the bays to drive in. Consideration should be given to the possibility of ceasing filling at an adjacent bay while the movement of another vehicle takes place.

Ensure road trucks ready to load are parked at a safe distance from the gantry and their movement is strictly controlled by the supervisor.

4.5 Loading gantries

Road tanker loading racks

Rail tanker loading racks

Gantries are structures that facilitate the loading of product into road and rail tankers from storage tanks. The structure houses loading arms and platforms for top loading tankers and flexible hoses for bottom loading tankers. Devices and instruments are provided for spill prevention and containment, and effective earthing/grounding of vehicles. Canopies provide rain protection for the product and operator.

A loading gantry should have:

- an adequate roof height and sufficient manoeuverable space for vehicles;
- sufficient vehicle-to-vehicle spacing for safe loading;
- adequate ventilation;
- adequate safety features on stairs and ladders, and fall protection systems;
- no trip hazards on platforms;
- adequate area lighting;
- emergency shower, eyewash station and emergency shutdown located in readily accessible locations;
- additional emergency shutdown button located away from the loading area;
- ready access to valves for testing;
- electrical equipment designed to meet the area classification drawing;
- fire protection and firefighting systems;
- good drainage that slopes away from the gantry into a slops recovery system;
- collection systems for drips.

Unsafe act: Risk of falling

Hydraulically actuated fall protection on road tanker loading gantry

Always ensure that:

- No ignition sources are brought onto the gantry.

- Routine inspection and maintenance is performed for the overfill protection system, loading arms and earthing cables.

- Loaders and operators are fully trained and adhere strictly to procedures.

- Procedures are updated following any engineering modifications.

- Engineering solutions are used to prevent abuse or misuse of equipment.

- The level of supervision provided is adequate.

- An instruction board is provided covering the safety critical tasks.

ACCIDENT **Major fire at a refinery road tanker loading gantry!!!**

Four vehicles were loading when the driver of one vehicle wedged open the 'deadman's' handle on the top loading arm with a wooden wedge and commenced loading motor spirit. For some unknown reason the loading arm came out of the compartment and covered the vehicle and the ground in motor spirit. It is thought that the vapour ignited on the vehicle exhaust pipe or the battery, causing a fire. The refinery fire brigade quickly extinguished the fire and no one was injured. The fire completely destroyed three road trucks and the gantry itself.

This and other forms of loading malpractice, i.e. jamming open the 'deadman's' valve handle and leaving vehicles unattended during loading, has been observed and reflects a need to educate drivers and loaders in loading procedures and the correct use of safety equipment. Supervision direct or remote, for example by TV cameras and audits, can identify and prevent these bad practices, but a technical solution should be aimed for so that overfilling, excessive loading rates etc. become less dependent on the driver's actions.

ACCIDENT **Road tanker damages fuel line causing fire at loading bay!!!**

An empty 22,500 litre (6,000 US gallon) semi-trailer road tanker was manoeuvring into position to load motor spirit at a loading bay when the vehicle's tank struck a valve on the motor spirit line. The valve sheared at the neck of the flange, releasing the line contents. Approximately 560 litres (150 US gallons) spilled over the front of the vehicle. When this happened, the driver immediately switched off the engine, which continued to run at an abnormally high speed. He jumped clear just before ignition occurred, and a severe fire ensued. The intense heat of the fire damaged other valves and flanges in this and the adjoining bay, causing more product to spill out. There was severe damage to the gantry and surrounding equipment.

This accident occurred due to the insufficient clearance between the gantry pipe system and the vehicle. The source of ignition was probably the diesel engine overspeed or the sheared off valve striking the battery box, splitting the fibreglass battery cover and arcing across the terminals.

4.6 Road car unloading

Before road trucks are offloaded at a facility, the following should be considered:

- A risk assessment to determine the necessity for unique couplings for each product/chemical or alternatively keeping the unloading point/valve locked.
- A hazard assessment (HAZOP) of the proposed unloading procedure and facility to check on the adequacy of safeguards.
- Sampling of the contents to ensure correct product and specification are in accordance with documentation.
- Assurance that the road truck is at the correct off-loading point.
- Clearly marked off-loading point with product name with NFPA/HAZCHEM labels.
- Clearly defined responsibilities for the activities undertaken by the driver and the receipt operator.
- Written unloading procedure.
- Competent drivers and receipt operators.
- Examination of the tank prior to unloading to check for any noticeable defects.
- Notice at receipt point detailing unloading procedure and correct Personal Protective Equipment to be worn.
- Emergency plans to cover all potential scenarios that could happen.

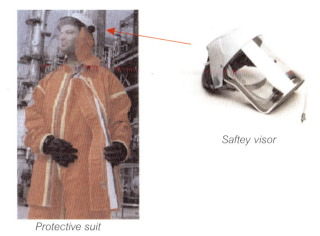

Saftey visor

Protective suit

Wear the correct Personal Protective Equipment when handling hazardous products and chemicals.

ACCIDENT A truck was unloading hydrochloric acid using compressed air when a leak occurred at the bottom of the tank car. The air pressure had created cracks in the protective internal lining and the acid found its way to the metal shell and quickly attacked it. The leak was plugged by the local fire brigade using special protective equipment under water curtains.

It is not recommended to use compressed air to unload road trucks. It is preferable to use a closed piping system and a pump.

ACCIDENT **Unloaded into the wrong tank!!!**

A facility, manufacturing fire retardants and water treatment chemicals, was expecting deliveries of epicholorohydrin and sodium chlorite by road truck on the same day. The two consignments, although originating from different countries, stayed at the same distribution centre for a period of time. At this stage, these two loads and their paperwork came together. But the paperwork was mistakenly switched and the sodium chlorite chemical container was transported across a number of European countries accompanied by documentation belonging to the tanker containing epichlorohydrin.

Site personnel commenced unloading operations upon arrival of the tanker, thought to contain epichlorohydrin, when a series of explosions ripped through the epichlorohydrin storage tank. Six site personnel were injured and five workers at adjacent facilities were injured by flying glass. The fire persisted for one hour and was brought under control by about 100 firefighters from across the area. A 100m (330 ft) plume of black smoke containing hydrogen chloride resulted in six firemen requiring extensive hospital treatment after breathing in the toxic fumes.

Unloading sodium chlorite into a tank containing epichlorohydrin was a recipe for disaster. (Sodium chlorite is a strong oxidizing agent and may interact and explode when mixed with organics or other oxidizable materials.) Oxidation of the epichlorohydrin took less than 90 seconds thereby minimizing the evolution of phosgene gas, a very toxic byproduct.

The oxidization of epichloro-hydrin caused a series of explo-sions and fire, releasing a 100m plume of black smoke containing hydrogen chloride.

Road tanker unloaded what was thought to be epichlorohy-drin but was actually sodium chlorite.

Storage tank containing epichlorohydrin received sodium chlorite by mistake.

Check that the right product will go into the right tank before unloading commences through sampling, documentation and labelling.

Good practices to avoid the hazards of misidentification of tanker contents are detailed in the publication *Procedures for offloading products into bulk storage at plants and terminals*, published by the Chemical Industries Association in 1999 (ISBN 1 85897 087 3).

ACCIDENT Overflow of road tanker while unloading under pressure!!!

Five minutes after commencing the discharge of a non-hazardous product from a tank truck (road tanker) into a shore tank using compressed air, the product flowed back from the shore tank into the tank truck and overflowed.

To transfer product from the tank trailer to the shore tank, three top manlids on the tank trailer are shut and an air hose is connected to a valve on top of the tank trailer, supplying compressed air at 1.8 bar (26 psig) from the customer, to push the product into the shore tank.

On the day of the incident, the driver had tightened the screws of all the manlids, except the first one, with a hammer. The first manlid (hatchcover) was only made hand tight and remained loose.

After connecting the product unloading hose from tank trailer to shore tank, the compressed air hose was connected onto the compressor line of the truck which runs to the top of the truck. Five minutes later, the product was coming out of the front top manhole of the truck.

Corrective actions

- All shore tank inlet lines will be fitted with non-return/check valves.

- Drivers will be instructed to remain in attendance during unloading operations and to monitor the pressure in the tank car.

- Training programs will be reviewed and updated.

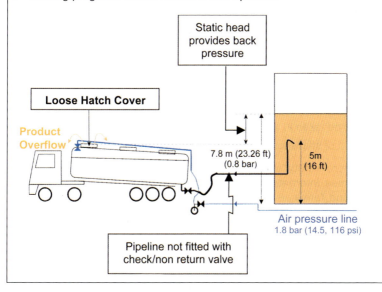

Static head provides back pressure

Loose Hatch Cover

Product Overflow

7.8 m (23.26 ft) (0.8 bar)

5m (16 ft)

Air pressure line 1.8 bar (14.5, 116 psi)

Pipeline not fitted with check/non return valve

5

Loading and unloading of ships

5.1 Introduction

Ships are used as the main mode of transportation of crude oil and hydrocarbon products in bulk quantities around the world. They receive their cargoes at tank terminals before making their way to their destinations to discharge their load. Loading and unloading of cargo is performed using loading arms and hoses. The concerns of inadequate or improper ship/shore interface and connection, surge pressures, stray currents, etc. have made the use of a ship/shore safety checklist an international industry recommendation.

5.2 Ship/shore safety checklist

Ship/shore activities are a joint responsibility of the ship's master and terminal (shore) representative. Before a tanker arrives at a terminal for loading or unloading, there should be a formal exchange of essential pre-arrival information between the tanker and the terminal to ensure all operations will be carried out safely in accordance with the port/terminal regulations. Prior to loading/discharging, the terminal representative and the ship's master must be satisfied that the procedures and arrangements for the safe transfer of cargo are fully acceptable to both parties. This includes:

- handling procedures, including maximum loading/discharging rates;
- actions to be taken in an emergency;
- completion of ship/shore safety checklist.

The ship/shore safety checklist (SSSCL) published by OCIMF/ ISGOTT (Oil Companies International Marine Forum/International Safety Guide for Oil Tankers and Terminals) covers the minimum checks and requirements prior to the commencement of ship/shore transfers.

It is completed jointly by a responsible ship's officer and a terminal representative. It is not a paper exercise. Each item on the checklist should be verified before it is ticked. This will entail several joint physical checks by the two responsible parties.

Does the terminal have a procedure in place to ensure that a pre-cargo transfer conference is undertaken, including completion of the ship/shore safety checklist?

Example of part of completed ship/shore safety checklist

	General	Ship	Ter-minal	Code	Remarks
1.	Is the ship securely moored?	✓	✓	**R** Every 2 hrs by terminal, constantly by ship	Stop cargo at 25 kts wind vel. Disconnect at 28 kts wind vel. Unberth at 35 kts wind vel.
2.	Are emergency towing wires correctly positioned?	✓	✓	**R** Same as mooring above	Approx. 1.5 m above water line at stern and bow.
3.	Is there safe access between ship and shore?	✓	✓	**R** Every hour	Fixed hydraulic gangway.
4.	Is the ship ready to move under its own power?	✓	✓	**PR** Every 2 hrs	Ship representative reports to jetty operator.
5.	Is there an effective deck watch in attendance on board and adequate supervision on the terminal and on the ship?	✓	✓	**R** Every 4 hrs	1 officer + 2 seamen on deck 1 shore representative at jetty
6.	Is the agreed ship/shore communication system operative?	✓	✓	**AR** Every 2 hrs	Provision of UHF radio by terminal as per terminal regulation. Regular communication check.
7.	Has the emergency signal to be used by the ship and shore been explained and understood?	✓	✓	**A**	Terminal's emergency regulations provided to ship. Horn and alarm are in accordance with terminal's regulation and location of terminal/jetty's alarm point given to ship.
8.	Have the procedures for cargo, bunker and ballast handling been agreed?	✓	✓	**AR** Every hour	Direct bunkering. Cargo plan and ballast plan submitted by ship to terminal.
9.	Have the hazards associated with toxic substances in the cargo being handled been identified and understood?	✓	✓		MSDS of product discussed with ship's representative.
10.	Has the emergency shutdown procedure been agreed?	✓	✓	**A**	Ship must not shut down emergency isolation valve before informing terminal representative at jetty. See terminal regulations.

ACCIDENT **Betelgeuse disaster!!!**

Fifty members of the crew and terminal staff on the island jetty were killed when the Betelgeuse broke up, exploded and caught fire during ballasting. The terminal was closed for the next 20 years.

The initiating event was the buckling of the ship's structure around the permanent ballast tanks. This was immediately followed by explosions in the tanks and breaking of the ship's back.

The initial failure was caused by:

- A seriously weakened hull due to inadequate maintenance (there was also excessive corrosion and wastage in the ballast tanks due to non-renewal of the cathodic protection).

- Incorrect ballasting that caused excessive stresses (the ship was not equipped with a loadicator/stress calculator for calculating bending moments and stressing forces at all states of loading or ballasting).

It should also be noted that at the time it was not a requirement for this ship to be fitted with an inert gas system. It is reasonably clear that the disaster would have been much smaller had the tanker been inerted. There was inadequate means of escape from the island berth in such an event and the fire water/foam system could not be initiated from the jetty.

It is current practice that:

- Ships are vetted to ensure no substandard ship is permitted at any berth.

- This type of ship must be fitted with an inert gas system in full working order (see *Section 5.12: Inert gas* on page 100).

- Loading/discharge plans are agreed between ship and shore representatives for cargo and ballasting to ensure stipulated stresses, stability and draft/trim of the vessel are not exceeded.

- Ships have a reliable stress calculator/loadicator.

- Jetties are protected with water sprays that can be automatically activated in the event of fire, and for island berths, equipped with enclosed lifeboats similar to those on offshore platforms.

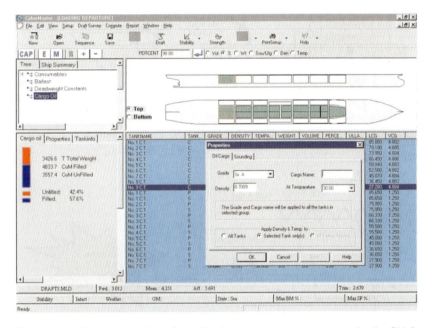

Stress calculation is typically performed using a computer programme by the Ship's Master/Loading Officer.

ACCIDENT Major gasoline spill and fire!!!

A tanker was discharging two leaded grades and one unleaded grade gasoline to a terminal. The ship had discharged the two leaded grades followed by a water plug. It was about 15 minutes into discharging the unleaded grade when an explosion occurred 100m (330 ft) off the starboard side of the vessel. A severe fire followed adjacent to the ship and burned for eight hours, injuring one person seriously and completely destroying the accommodation area. The spill, estimated at 3,000 tons of gasoline, was caused by the failure of the ship to close the sea chest valves following the completion of the water plug. The most likely ignition source was a passing fishing vessel.

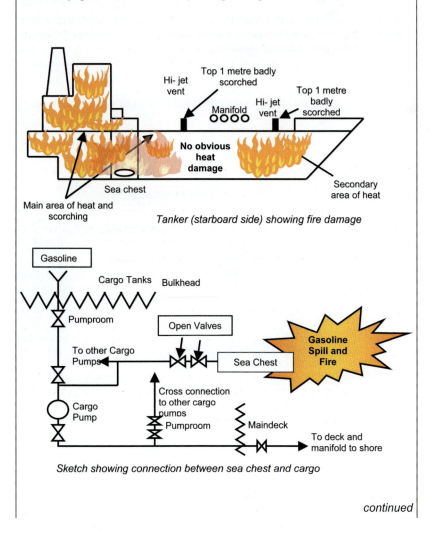

Tanker (starboard side) showing fire damage

Sketch showing connection between sea chest and cargo

continued

Lessons learned

- Instructions should be issued regarding the sequence of operations during critical changeover from cargo pumping to ballasting/operation of sea chest valves. These instructions should be issued by the tanker's owner and included in the operations manual onboard.

- Clear instructions should be assured under the vessel's ISM Code defining the responsibility for the operation of overboard and sea chest valves. Responsible personnel on board must be thoroughly competent to follow these procedures.

- The requirement for latching and sealing sea chest valves in the closed position must be emphasized in operating procedures and ship/shore checklists. Both terminal and ship representatives should cooperate in checking the sea valves which should have positive and reliable indicators showing the position of the valve when shut, fully open or at any position in between.

5.3 Ship/shore connections

Ships are loaded/discharged using either hoses or marine loading arms.

The ship's manifold must be kept within the safe operating envelope of the loading arm or hose string during the loading/discharging operations. The ship's moorings must therefore be continually monitored to ensure that movement is restricted to within this safe operating envelope.

5.4 Hoses

Hoses should conform to a recognized standard for the specific material to be handled. Each length should be marked by the manufacturer in accordance with ISGOTT (International Safety Guide for Oil Tankers and Terminals) requirements and have a corrresponding individual life sheet.

Hoses must be inspected before use for defects, handled with care with bridles and saddles and tested periodically in accordance with the OCIMF (Oil Companies International Marine Forum) publication *Guide for handling, storage, inspection and testing of hoses in the field*.

The maximum permissible working pressure and flow rate must not be exceeded and surge pressures must be avoided.

To avoid stray currents flowing through hose systems between ship and shore, either one length of the hose string should be non-conducting (insulating) or an insulating flange should be fitted at the connection of the hose string to the pipeline system. (Concern about stray currents is quite separate from static electricity—see *Section 3.1: Static electricity* on page 35.)

Good practice

Safe access from shore to ship for terminal representative and ship's crew

Well supported hose during lifting

Hydraulic test apparatus for proving integrity of hoses at periodic intervals

Poor practice

Use of steel wires in direct contact with hose should not be permitted. Bent to a radius less than that recommended by the manufacturer.

Flexible hoses

The main concerns identified from audits are as follows:

- no blind end flange or missing bolts;

- damage from handling;

- no retirement criteria and no maximum life criteria;

Damage from handling

- no annual hydraulic test;

- no immediate destruction of retired hoses;

- bad storage conditions for new hoses;

- no insulating flange (see Section 5.8 on *Stray currents and insulating flanges* on page 90);

- poor alignment of hose/boom to ship's manifold.

Poor practice

Messy hose storage

New flexible hose stored in a bund area without end flanges and exposed to bad weather conditions.

Leaking hoses

Note: Kinked hose and poor alignment to ship's manifold

Missing bolts

Missing bolts

Gasoline leak   *Damage from handling*

89

5.5 Loading arms

Articulated metal loading arms are used instead of hoses for transferring products to or from ships with the capability of accommodating differences in tides, freeboard and ship motions.

They are designed to cater for a specific safe operating envelope.

A loading arm

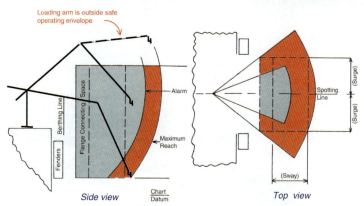

Safe operating envelope of loading arms (Taken from design and construction specification for marine loading arms, OCIMF)

Poor practice

Result of loading arm going outside its safe operating envelope.

When the size of the loading arm is different from that of the ship's manifold, a spool piece will be used at the point of connection (shown in the sketch below). Ensure the correct specification is checked by ship and terminal representatives.

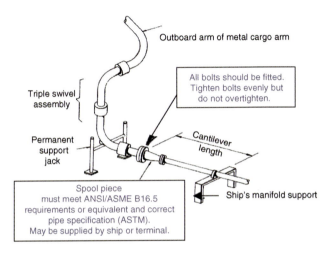

Outboard arm of metal cargo arm

All bolts should be fitted. Tighten bolts evenly but do not overtighten.

Triple swivel assembly

Permanent support jack

Cantilever length

Spool piece must meet ANSI/ASME B16.5 requirements or equivalent and correct pipe specification (ASTM). May be supplied by ship or terminal.

Ship's manifold support

5.6 Spheres/pig launchers/receivers

Berths are often equipped with sphere launchers/receivers in order to clear long lengths of piping to and from a jetty and storage tank of a previous product to prevent contamination.

These operations demand careful attention to prevent accidents. Reference should be made to BP Process Safety Booklet *Hazards of Trapped Pressure and Vacuum* for additional guidance on this subject.

5.7 Emergency shutdown (ESD systems)

An emergency shutdown procedure must be agreed between the ship and the terminal and recorded on an appropriate form. It may include a special device for the emergency disconnection of cargo hoses or loading arms (see sections 5.10 and 5.11 on page 95 and 98).

An ESD stop button should be located adjacent to a jetty escape route.

Difficult to see the ESD button and no indication of what is shut down. ESD button with large and clear signage.

5.8 Stray currents and insulating flanges

Large currents can flow electrically between ship to shore and vice versa. An all-metal loading arm provides a very low resistance connection between ship and shore and there is a very real danger of an incendive arc when the ensuing large current is suddenly interrupted during the connecting or disconnecting of the hose/arm at the tanker manifold. If stray currents are present in berth piping, connecting and disconnecting of a ship's hose may produce arcs because the resistance of the ship's hull to ground (via seawater) is exceedingly low.

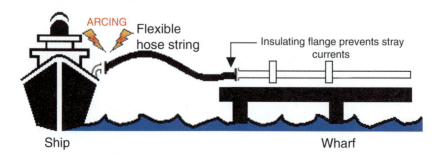

An insulating flange (IF) is a flanged joint incorporating an insulating gasket, sleeves and washers to prevent electrical continuity between pipelines, hose strings or loading arms.

- Each loading arm or hose string (and vapour recovery connection, if fitted) must be insulated in accordance with ISGOTT.

- The insulating flanges must not be painted, greased or damaged.

- Insulating flanges should be tested for resistance every six months. An IF should provide resistance greater than 1,000 ohms.

- Connecting flanges in the hose string must be supported clear of the berth structure to prevent insulating flange being rendered ineffective.

> **A ship/shore bonding cable is not effective as a safety device and may even be dangerous.**
>
> **A ship/shore bonding cable should therefore not be used.**

Ship-to-shore bonding wire

ISGOTT does not recommend a bonding wire between ship and shore. The bonding wire has no relevance to electrostatic charging. Its purpose was to attempt to short circuit the ship/shore electrolytic cathodic protection systems so that currents in hoses and metal arms would be negligible. It has been found to be quite ineffective and a possible hazard to safety. Insulation flanges or a single length of non-conducting hose is recommended to prevent the flow of current between the ship and shore. Refer to ISGOTT for more details.

Although the potential dangers of using a ship/shore bonding cable are widely recognized, attention is drawn to the fact that some national and local regulations may still require a bonding cable to be connected. If a bonding cable is insisted upon, it should first be inspected to see that it is mechanically and electrically sound. The connection point for the cable should be well clear of the manifold area. There should always be a switch on the jetty in series with the bonding cable and of a type suitable for use in a Zone 1 hazardous area.

It is important to ensure that the switch is always in the 'off' position before connecting or disconnecting the cable. Only when the cable is properly fixed and in good contact with the ship should the switch be closed. The cable should be attached before the cargo hoses are connected and removed only after the hoses have been disconnected.

> **Use an insulating flange or a single length of non-conducting hose to ensure electrical discontinuity between ship and shore, as required by ISGOTT.**

ACCIDENT **Electrical spark at ship/shore loading arm connection!!!**

A tanker was berthed, awaiting the loading of crude oil. During the procedure of connecting up the loading arms, a technician bypassed the special insulating flange located in the loading arm system with a strap/cable. This allowed a continuous path for the flow of electrical current between ship and shore. As the berth's vapour return loading arm came into close contact with the tanker's pipe manifold on deck, a spark occurred between the flange faces.

The confusion between guarding against static electricity and preventing stray currents probably lead to the technician's actions. While static charge generation is a concern during loading of product, stray currents can occur while connecting up and disconnecting the loading arms.

Stray currents and static electricity are different hazards that require different preventative measures.

Ship/shore insulating joints/flanges in loading arms or sections of non-conducting hose are critical safety items that must be identified, tested and maintained to secure their continuing integrity in accordance with ISGOTT requirements.

Insulating flanges with good signage indicating their location to prevent inadvertent removal or painting.

ACCIDENT **Resistance of ship/shore insulation flanges destroyed by painting!!!**

During a scheduled inspection, the insulation flange joints on the marine loading arms were discovered to have inadequate resistance to prevent a ship to shore current flow. The insulation joints were inadvertently painted when the loading arms were painted. This would have provided a continuous path for current flow. Arcing could have occurred during any connecting/ disconnecting of a loading arm to a ship's manifold.

5.9 Surge pressures

A pressure surge is generated in a pipeline when there is a sudden change in the liquid flow rate, such as when a ship's ESD (Emergency Shutdown) valve closes rapidly, the liquid immediately upstream of the valve is brought to rest almost instantaneously.

Common causes of high pressure surges are:

- closure of an automatic emergency shutdown valve;
- slamming shut of a non-return/check/butterfly type valve;
- rapid closure or opening of a manual or power-operated valve;
- starting or stopping of a pump.

It is possible for pumps and valves to produce pressure surges in a pipeline system. These surges may be sufficiently severe to damage the pipeline, hoses or loading arms. One of the most vulnerable parts of the system is the ship to shore connection.

Pressure surges are produced upstream of a closing valve and may become excessive if the valve is closed too quickly. They are more likely to be severe where long pipelines and high flow rates are involved.

If the pressure stresses or displacement stresses resulting from a pressure surge exceeds the strength of any affected part of the piping system, there may be leakage or rupture leading to an extensive spill of oil.

Butterfly and pinned back non-return valves in ship and shore cargo systems have been known to slam shut when cargo is flowing through them at high rates, thereby setting up very large surge pressures which can cause line, hose, or loading arm failures and even structural damage to jetties.

> **The magnitude of a pressure surge depends on the closure time of the valve that in turn depends upon the design of the valve.**

To avoid pressure surges, valves at the downstream end of a pipeline system should, as a general rule, not be closed against the flow of liquid except in an emergency.

Although cargo transfer systems can be designed to be inherently safe for a surge pressure, a compromise with good practice must be achieved.

ANSI B31.4 specifies surge calculations be carried out and protective measures be provided, with an overpressure allowance of 10%.

95

How to alleviate surge pressures?

- Linked ship/shore ESD systems, with correct shutdown sequence* irrespective of whether ESD is initiated on the ship or on shore, will reduce the potential surge pressures on ESD. (*Terminal shuts down first at loading whilst ship shuts down first on discharging).

- Reduce loading/discharge rates to a safe level for the corresponding Total Valve Closure Times.

- Set the isolation valve closure times to avoid surge pressures.

- Ensure ship/shore checklist covers if necessary the need to shut down the shore transfer system before the ship closes its isolation valve against a full flow rate.

ACCIDENT **Sudden closure of butterfly valve causes surge pressure!!!**

During the discharge of crude oil cargo from a ship to terminal storage, a butterfly valve at the base of the marine loading arm failed causing it to 'slam shut' against the flow of oil. This sudden closure caused a pressure surge resulting in two of the ship's 24-inch (600 mm) cargo lines being breached in way of expansion couplings.

Damage to ship's 24-inch (600 mm) pipework in way of expansion couplings

Damage to Marine Arm's 16-inch (400 mm) butterfly valve

ACCIDENT **Bellows fail due to surge pressure!!!**

An export pump was started with its discharge valve fully open and the suction valve at the tank closed. The resulting pressure surge wave caused the tie-bar mounted lugs on the suction line bellows to fail, ripping open the outlet pipe from a storage tank. This resulted in the spillage of 50,000 litres (13,200 US gallons) of ultra low sulphur petrol/gasoline, which was contained within the tank bund/dike. A layer of foam was applied to the spill area to prevent ignition.

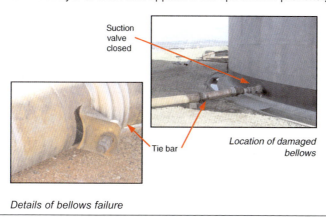

Suction valve closed

Tie bar

Location of damaged bellows

Details of bellows failure

5.10 Emergency release coupling (ERC)

ERC is also termed Dry Break Coupling. The ERC where fitted forms an important part of an emergency release system for marine loading arms, and is designed to safely disconnect the arm from the ship with minimum spillage in an emergency.

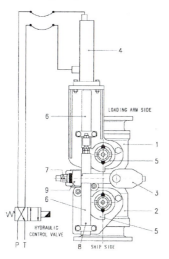

Schematic of typical ERC

1) Ball valve loading arm side

2) Ball valve ship side

3) Clamping system

4) Double acting hydraulic cylinder

5) Ball valve actuating lever

6) Push rod system

7) Lock nut

8) Shear pin (ball valve)

9) Shear pin (disconnection)

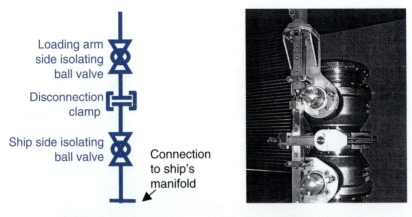

Loading arm
side isolating
ball valve

Disconnection
clamp

Ship side isolating
ball valve

Connection
to ship's
manifold

Main elements of an ERC

A picture of an ERC
Taken from Schwelm Verladetechnik GmbH

The ERC as shown in the diagram above is operated by one hydraulic cylinder that closes the ball valves and subsequently separates the clamp connection.

The most important feature of the ERC is to ensure full closure of the valves before disconnection. Special tools are normally required to re-open and reassemble the ERC after disconnection.

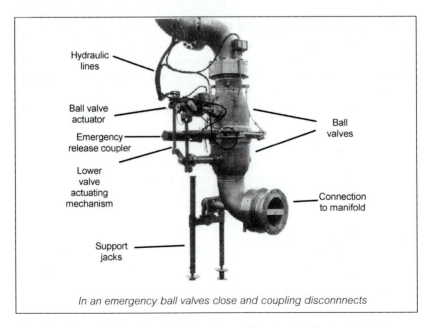

Hydraulic
lines

Ball valve
actuator

Emergency
release coupler

Lower
valve
actuating
mechanism

Support
jacks

Ball
valves

Connection
to manifold

In an emergency ball valves close and coupling disconnnects

Taken from Liquefied Gas Handling Principles on Ships and in Terminals, McGuire and White

Good practice

- The design of the system must undergo a Failure Mode and Effects Analysis to prove inherent integrity.

- The ERC mechanism should be protected from interference by jetty personnel.

- The staff involved in operation and maintenance of loading arms need to understand exactly how the ERC works.

- The system should not be used in normal operations to shut off flow.

- Functional tests where the ball valves shut and the ERC is stroked should be performed prior to each shipment.

- Full ERC separation tests should be performed annually.

- No maintenance or troubleshooting of the ERC should be carried out unless it is disarmed and bolted with safety bolts.

- Maintenance/troubleshooting should be authorized through the Permit-to-Work system.

ACCIDENT **Failure of emergency release coupling at jetty!!!**

An ERC inadvertently activated disconnecting the loading arm without prior closing of the isolating valves, resulting in the release of approximately 8.5 tons of propylene which fortunately did not ignite.

View as propylene cloud develops. The aft of the LPG carrier is the only apparent part of the vessel.

ERC must only be released if isolating ball valves are SHUT first. The integrity of the interlock mechanism must be guaranteed.

Lessons learned

- Operating and maintenance manuals on ERCs should identify all *safety critical items* with their inspection, maintenance and testing requirements.

- Competency assessment should be an integral part of any quality training and refresher training program for jetty operators.

- Location and testing of ESD at jetties is critical for a quick and reliable effective response to an emergency.

5.11 Quick connect/disconnect couplers (QC/DC)

A QC/DC is either a manually or hydraulically operated device used to clamp the loading arm flange to the ship's manifold without the use of bolts. The QC/DC must be provided with an interlock to prevent any possibility of an inadvertent release.

All QC/DC devices should be designed, tested and maintained in accordance with OCIMF specification for marine loading arms.

A QC/DC and ERC located on a loading arm

ACCIDENT Failure of ship/shore couplers(QC/DC)–Case 1!!!

A component within a ship/shore quick connect/disconnect (QCDC) coupler failed during operation, causing a coil spring to eject violently.

The key lessons are as follows:

- Pre-load on the QC/DC coupler springs could be accidentally altered, increasing the operating load on the knuckle joints to an unacceptable degree, if the grub screws that fix them to the spring struts become loose.

- Some ships may carry thicker flanges that could exceed the operating tolerance of the coupling. Jetty operators need a simple template to check the thickness and diameter of flanges on ship's manifold before attempting to operate a QC/DC device.

- Gaskets should not be used on the joint as QC/DC devices are generally designed to seal metal to metal with a fixed 'O'-ring insert.

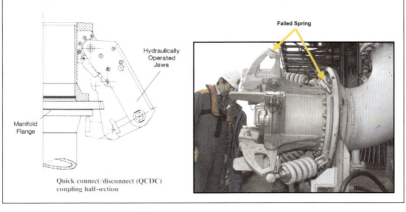

ACCIDENT Failure of ship/shore couplers–Case 2!!!

The ram shaft of a coupler failed when the pivot point seized—the shaft being subjected to extreme pressure when it was in the open position. All moving parts require periodic lubrication (greased).

5.12 Inert gas

Where ships are required to be fitted with operational inert gas or nitrogen systems, the following conditions must be maintained:

- Closed gauging systems that are normally fitted with vapour locks to permit sampling without reducing the inert gas pressure.

- Positive pressure in the cargo tanks and slops tanks to prevent any air being drawn in. Failure of the inert gas system requires immediate attention by the ship's master.

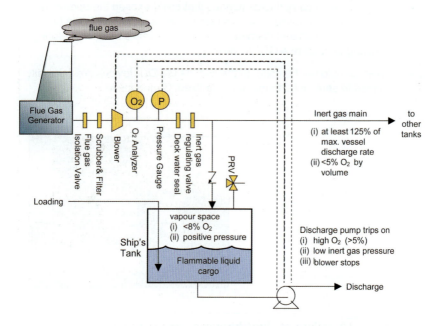

General layout of inert gas system on a ship

Inert gas systems must remain fully operational during the discharge of cargo.
When a ship is fitted with an Inert gas system, it should be used, whatever the product being handled.

ACCIDENT Crew member is overcome by inert gas and H₂S when opening gauge hatch!!!

A ship was discharging with the cargo tanks inerted and under positive pressure. A crew member was on the ship's deck engaged in cleaning the sensor on a cargo tank measurement gauge. Some five minutes after going to attend to the task, the crew member was found dead beside a tank gauge.

The cargo tank was fitted with a vapour lock arrangement to facilitate cleaning of the sensor but this was not used. Instead an inspection plate had been removed from the side of the gauge exposing the deceased to a high concentration of H_2S and N_2 from the cargo vapour space.

Cover
Plate
Removed

The measurement cargo gauge on the ship

There was a failure to transmit the H_2S hazard of the cargo and precautionary measures through procedures such as the ship/shore safety checklist at loading and discharge ports. However, by breaking containment, the crew member was also exposed to a local environment deficient in oxygen. Exposure to an atmosphere containing less than 10% oxygen content by volume inevitably causes unconsciousness irrespective of the H_2S content. Death can result unless the victim is removed to the open air and resuscitated quickly. Refer to BP Process Safety Booklet *Hazards of Nitrogen and Catalyst Handling*.

Pyrophoric iron sulphide is formed when H_2S reacts with rusted steel in the absence of oxygen. Iron sulphide deposits will heat and autoignite when coming into contact with air. Crude oil ships must therefore ensure that a failed inert gas system is repaired and restarted before discharge or deballasting is allowed to continue.

6

Tank level measurement and overfill protection

6.1 Level measurement

Accurate level measurement/gauging is required to determine the inventory for customs purposes, to ensure that product losses such as tank leaks are not occurring, and for preventing overfills.

Gauging can be performed automatically or manually. Automatic measurement uses level indication devices while manual measurement is done using tapes/rods.

Level gauge on a tank roof

Radar level gauge

Level indicatiors

Tank overfills will result in environmental and safety hazards with the potential for major fires and explosions.

Overfilling can also cause damage to the sealing ring of a floating roof. Also, when the roof is at a level above the top of the tank, firefighting foam cannot be contained between the shell and the foam dam. Product spillage causes an even more serious fire hazard.

Dip tape

ACCIDENT Major tank fire from tank overfill!!!

A gasoline tank overflowed and the vapours entered the nearby buildings that included the fire station, medical centre, maintenance workshop, store, engineering office and administrative building. Eight people were killed and 13 hospitalized.

It is not known how long the tank was overflowing before a security guard detected a strong smell and notified shift control. Two operators were sent to investigate and it is believed that the vehicle they were driving ignited the vapours.

The resultant blasts damaged parked cars, fire engines and shattered windows in nearby company buildings and houses. These were sufficient to rattle windows 20 km (12 miles) away.

A resultant full surface fire on the 1.5 million litres (9400 bbls) gasoline tank spread to four other larger tanks before it was contained.

It is understood that the level instrumentation and high level alarm may have been unreliable.

Blast damage to vehicle

Full surface tank fires

The Buncefield incident shortly described below is another example of process equipment overfilling that escalated to a major incident.

ACCIDENT Around 19.00 on 10 December 2005 Tank 912 at the Hertfordshire Oil Storage Limited West site started receiving unleaded gasoline. From around 03.00 on 11 December 2005 the Automatic Tank Gauging (ATG) system for Tank 912 recorded an unchanged reading but filling continued. The ATG system enabled the operator to monitor levels, temperatures and tank valve positions, and to initiate the remote operation of valves all from the control room.

At around 05.20 Tank 912 would have been full and starting to overflow through the roof vents. No signal was received from the independent ultimate high level alarm protection system to automatically close off valves. Approximately 300 tonnes of gasoline, rich in butane as it was winter grade, spilled into the bund.

The overflow from the tank led to the rapid formation of a rich mixture of fuel and air. The vapour started to spread in all directions and thickened to about 2m deep. The vapour cloud ignited at 06:01 in a devastating explosion that reached 2.4 on Richter scale. The exact ignition source is not known at this time.

The terminal suffered extensive damage along with adjacent buildings and another oil terminal operated by the British Pipeline Agency. More than 40 people were injured and the following fire destroyed more than 20 tanks over three days, destroying the Hertfordshire Oil Storage owned and British Pipeline Agency operated terminals.

View of terminal damage 24 hours after the explosion

The UK HSE reports on Buncefield are available on the following web-site: http://www.buncefieldinvestigation.gov.uk/index.htm

To read more on tank fires, refer to the BP Process Safety Booklet *Liquid Hydrocarbon Tank Fires: Prevention and Response* ISBN 0 85295 504 9.

6.2 Manual gauging

'Hand-dipping' using a calibrated tape requires certain precautions to be taken. Refer to section 3.1.10 *Sampling, dipping and taking temperatures* for more information.

6.3 Automatic gauging

Level indicators can be provided with local readout and remote readout. Limit switches for high, low and intermediate level alarms can be included as part of the automatic level gauge with alarm signals transmitted to the control room. Low level switches and alarms may be necessary for:

- switching off mixers when the floating roof approaches the low operating position.

- stopping product withdrawal before the roof lands on its supports in the low position.

High level alarms are recommended to give warning of the roof approaching its maximum filling level and high-high level alarms are used to automatically switch off the pumps. An additional independent level switch for the high level alarm function is recommended. This can be a simple but very reliable switch that is triggered by the movement of the roof. Reliability of such instruments and overfill protection should be carried out based upon an acceptable code, such as IEC.61508.

Independent high level alarm on a floating roof tank

6.4 Tank level settings for alarms

(Refer to API RP 2350)

The figure below illustrates an example of the alarm settings that may be used. The actual times may be longer or shorter, depending on operating practices.

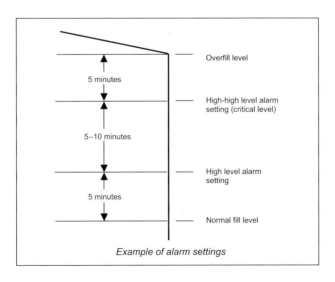

Example of alarm settings

107

ACCIDENT **False reading causes tank overflow!!!**

A level gauge became stuck and gave a false level reading while receiving an import of atmospheric residue at a refinery. The gauge was not installed correctly. When the ship's discharge rate appeared too slow (the gauge had stuck), the boardman assumed that the ship was nearing the end of its discharge and therefore did not communicate with the jetty operator. As a result, the tank was overfilled and $100\,m^3$ (26400 gallons) of residue overflowed into the bunded area, of which $85\,m^3$ (22450 gallons) were recovered.

Improved liaison between operators in monitoring transfers of hydrocarbons is required to avoid overfilling tanks and information on unreliability of instrumentation needs to be communicated between shifts and immediately rectified.

ACCIDENT **Gauging system failure leads to overfilling of crude oil tank!!!**

During the transfer of crude oil into a storage tank, the tank overflowed spilling approximately 3200–4800 litres (20–30 bbls) of crude oil. A lighting fixture was damaged by the abnormal height of the floating roof. Fortunately, this did not ignite the spillage. Investigations showed that the tank's remote gauging system had malfunctioned some hours previously but this was not detected. The High Level Alarm was inoperative, apparently as a result of previous modifications made to the gauging system.

Fortunately, the shift team had calculated the time to fully fill the tank and was investigating the situation as oil began to come over the rim. As a result of this added precaution, the spill volume was small. The tank level was restored to normal by transferring oil to another tank.

This incident demonstrates the importance of maintaining extra high level alarms and ensuring regular testing is carried out to confirm the reliability of all level instrumentation. This level of protection should not be compromised by any modification work carried out on the gauging system.

After any modification, perform a live test on the gauging system (on line) to ensure it works as intended when required. Include a simple procedure in tank-filling Standard Operating Procedures to calculate the estimated time to fully fill the tank up to the top dip, based on filling rates, as a backup to the tank gauging equipment. Preferably, install an independent High-High Level Alarm and Emergency Shutdown System, which automatically stops the feed pump on a High-High Level Alarm.

> **Ensure level instrumentation and alarms are tested at stipulated intervals to guarantee their continued integrity.**

108

6.5 Overfill protection

The following safeguards should be considered to prevent overfilling of tanks during transfers:

- At the anticipated flowrate, calculate the time when the tank is expected to be full.

- Gauge tanks at frequent intervals during product receipt and record rate of change in level over time.

- Equip tanks with high-level detection devices independent of any tank gauging equipment that will automatically shut down or divert flow.

- Design high level alarms and independent high level alarms to provide sufficient time for an orderly shutdown or diversion of product before the tank overfills.

- Maintain frequent communication with the supplier so that flow can be promptly shut down or diverted.

- Locate alarms where personnel who are on duty throughout product transfer can promptly shut down or divert the flow.

- Prepare instructions covering methods to check for proper line-up and receipt of initial delivery for tank designated to receive shipment.

- Check that *line up* of valves is *correct* prior to product transfer.

- Post warning signs around the tanks, listing the potential hazards and prohibit simultaneous tank filling and fuel dispensing.

- Provide training and monitor the performance of operating personnel.

- Prepare schedules and procedures for inspection and testing of gauging equipment, high-level instrumentation and related systems.

- Check and calibrate level instrumentation and alarms periodically to ensure their continual integrity.

- Investigate immediately any reports of spurious high level alarms.

ACCIDENT Incorrect line-up causes tank overflow!!!

During a routine transfer of lubricating oil from one tank to another, a third tank overflowed, spilling 247 tonnes into the bund/dike and piperacks. The immediate cause of the incident was the inlet valve to the third tank that was left open following completion of a previous transfer. Contributory causes were the failure of the gauge to show the true level and inadequate monitoring arrangements for changes in tank levels. The inaccurate level reading prevented an alarm since it came from the same level measuring device.

The standard operating procedure did not cover the correct line up of valves before the commencement of product transfers.

7

Tank inspection and maintenance

7.1 Typical damage mechanisms

Typical damage mechanisms of storage tanks
• Internal corrosion
• External corrosion (shell, roof, base plate)
• Corrosion under insulation (CUI)
• Bimetallic corrosion—storage tank earthing system
• Tank settlement

• Structural failures:
  ○ jamming of rolling ladder preventing free movement;
  ○ loss of buoyancy on floaters;
  ○ pressure/vacuum valve or vent blockage causing roof/shell failure;
  ○ roof tilting on floaters due to accumulation of water/product.

• Combination of the above

Refer to **API 653** and **EEMUA 159** for further information.

7.2 Corrosion of tanks

Most tanks are made of carbon steel, which can corrode when exposed to air and water. Over time, uncontrolled rusting can weaken or destroy the components of a tank, resulting in holes or possible structural failure, and release of stored products into the environment.

Rusting is accelerated by factors including:

- increased temperature;
- corrosive environment;
- stray electric currents between interconnected components.

Corrosion is a major contributor to tank failure. Modern corrosion control combines historically proven methods with state-of-the-art technology such as cathodic protection to prevent tanks from deteriorating.

The following corrosion control strategies are common and can be used individually or in combination:

- use of corrosion resistant materials for constructing tanks;
- application of coatings and/or linings (paint, plastic, fibreglass, etc.);
- various forms of cathodic protection to prevent deterioration of tank components in contact with soil; and
- use of inhibiting chemicals in stored product to control corrosion of tank interior.

Corrosion product/debris collected on roof

Rust, a product of steel corrosion, can restrict the flow of water to drainage sumps, damage the roof seals and block drainage slots at the bottom of foam dams allowing water to accumulate, resulting in corrosion behind the dam. The result of excessive thinning of the shell plates from corrosion is seen below.

Failed/buckled tank shell due to excessive internal corrosion

ACCIDENT Tank collapsed due to corrosion under insulation!!!

A 5,000 m³ (1.3 million US gallon) insulated tank at a chemical terminal was emptied when corrosion holes were found on the roof. The tank was badly corroded. A severe gale, worsened by being funnelled between other tanks, caused the rusty tank to collapse.

Tank collapsed due to rustiness and severe gale

ACCIDENT Tank buckles in the wind!!!

A rare high wind condition combined with local thinning/pitting on the tank shell caused buckling. API 650 requires that the shell of a storage tank be checked for buckling and stability against overturning due to high wind load.

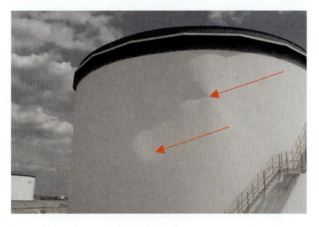

Buckled floating roof tank

ACCIDENT

A failure in the tank roof area occurred, causing crude oil to accumulate on the tank roof. Crude oil odours generated some concern in the community, including two nearby schools, that drew the attention of environmental regulators.

7.3 General inspection techniques

Generally, there are three types of inspection for storage tanks:

- visual, external inspection by operator;
- in-service inspection;
- out-of-service inspection.

External and internal inspections must be undertaken at regular intervals by an in-house or external 'Inspection Authority' in order to confirm the tank's continued integrity and safety. All the different types of inspection and inspection checklists are available in **API Standard 653 'Tank Inspection, Repair, Alteration and Reconstruction'**.

'Inspection Authority' = an individual or group that is qualified as an inspection specialist

Safety is enhanced by accomplishing the following:

- An 'Inspection Authority' should formulate an inspection program (both internal and external (in-service) examinations) for every tank and its associated pipework, instrumentation, relief arrangements, etc.
- Inspection reports form the basis for any management decision related to type and frequency of inspections and when a tank should be taken out of service. Inspection intervals should not be changed without approval through a 'Management of Change' procedure.
- The 'Operator' should track that reported defects and/or deficiencies are corrected within a stipulated timeframe.

- In addition to periodic internal and external inspections by the 'Inspection Authority', the 'Operator' of each storage tank should undertake a periodic visual check of critical items using a prepared checklist.

7.4 Inspection intervals

The interval between inspections of a tank (both internal and external) should be determined by:

- service history of a given tank or a tank in similar service;
- the corrosion allowances and corrosion rates of the tank;
- nature of products;
- results of visual maintenance checks;
- corrosion prevention systems in place;
- condition at previous inspections;
- location of tanks, such as those in isolated high risk areas;
- change in operating mode (for example, frequent floating roof landing);
- National Standards.

Operation beyond the time when a planned inspection is due can only be permitted by management after the risk(s) has been assessed through a formal 'Management of Change' procedure.

7.5 Visual inspection by operator

The external condition of the tank should be monitored by close visual inspection on a routine basis. The owner or any operator may perform the inspection. Personnel performing this inspection should be knowledgeable of the storage facility operations, the tank, and the characteristics of the product stored.

This routine inspection shall include a visual examination of the tank's exterior surface checking for:

- leaks;
- shell distortions;
- signs of settlement;
- corrosion; and
- condition of the foundation, paint coatings, insulation systems and appearance.

Inspections are carried out by the 'Operator' at more frequent intervals than those carried out by the 'Inspection Authority'. Intervals between these checks may vary from once per shift to once every six months depending upon the device/item.

> **Critical items associated with the safe operation of storage tanks should be formally checked by the Operator on a regular basis.**

Critical items that should be covered on a checklist by the Operations Department are as follows:

Critical items for inspection	
1) Vents and pressure/vacuum valve	
2) Tank shell	Fixed and floating roof tanks
3) Bund/dike drain valves	
4) Grounding/earthing equipment	
5) Tank top access, walkways, emergency exits, etc.	
6) Floating roof	
7) Roof drain valves	
8) Roof drain sump	
9) Pontoon compartments (including LEL testing)	
10) Seal, weathershield and metallic shunts	Floating roof tanks
11) Lower internal roof drain valve	
12) Emergency roof drains	
13) Wax scrapers	
14) Guide/gauge poles	
15) Ladders	
16) Foam pourers and foam dam	

These checks are detailed below. Records should be kept.

1) Check *vents and pressure/vacuum valve* screens on tank roofs to ensure they are clear and are fitted with the correct size mesh.

Vent blocked with wax

2) Check the bottom of the *tank shell* for settlement, vegetation, corrosion and signs of leaks.

Oil spill from leaking roof drain

3) Check *bund/dike drain valves* to ensure that they are kept *SHUT* to collect any spilled or leaking product. These may be opened to drain rainwater under strict supervision. Post a notice to this effect adjacent to all bund/dike drain valves. It should be noted that standing water left in bunds/dikes will increase the rate of corrosion of equipment and tank bottoms.

> **Keep bund/dike drain valves SHUT to collect any spilled or leaking product, except when draining rainwater.**

4) Check that *grounding/earthing* straps, cables and stainless steel shunts are in sound condition. Poor electrical paths increase the risk of ignition in seal covers at times of thunderstorms/lightning.

Tank grounding system components such as shunts or mechanical connections of cables shall be visually checked.

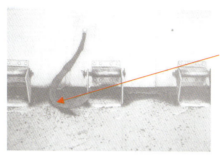

Grounding cable between weather shield and deck of a floating roof that is not maintained in good condition

5) Ensure that *tank top access, walkways, emergency exits, etc.* are in good condition. Prevent and remove slip and trip hazards in the vicinity.

ACCIDENT **Sampler nearly falls to his death!!!**

An operator was on a cone roof tank during icy conditions when he slipped and fell. He slid along the roof and saved his life only by grabbing the roof handrail.

---

6) Look for signs of oil on the *floating roof*, indicating small holes and cracks in welds in the deck of floating roofs which could ultimately lead to the roof sinking or catching fire.

Example of roof showing product from cracks

On floating roof tanks that have a seal mechanism with pantograph hangers, look for vertical grooves in the shell as a sign of a damage mechanism.

Check secondary seals are not folding down, as that indicates excessive gaps or movement of the roof.

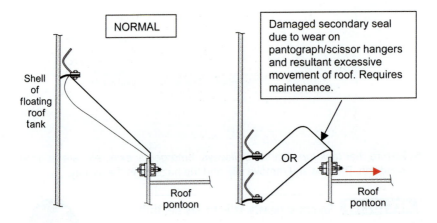

7) Check that the *roof drain valve* is kept open. Water on a floating roof is directed from the roof via a flexible hose, or a metal pipe with articulated joints, to an outside gate valve near the bottom of the shell. This valve should be kept *OPEN* at all times. Although there will be a risk of the tank contents emptying into the dike/bund should a complete failure occur to the internal hose or articulated pipe, the likelihood of this event is considered lower than the chances of someone forgetting to open the valve after sudden, heavy rainfall, which could cause the roof to sink. If there is standing water on the roof, check that it has not caused the tank roof to tilt or hang up. This drain should be checked once per shift to detect any leakage from the hose or articulated joints. If leakage is discovered, close the valve and report the leak to the area manager immediately.

> **Keep roof drain valve OPEN to prevent floating roof from sinking during sudden, heavy rainfall.**

8) Check the *drain sump* covers/screens and the check (non-return) valve that handles the flow of rainwater on floating roofs to ensure that they are not blocked with debris or wax and that the check valve/flapper is free.

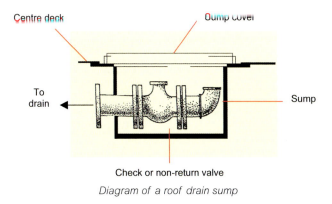

Centre deck

Dump cover

To drain

Sump

Check or non-return valve

Diagram of a roof drain sump

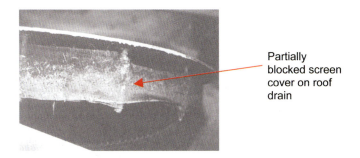

Partially blocked screen cover on roof drain

The rainwater collection sumps on the tank floating roofs are fitted with screens to prevent debris, scale, etc., from entering the drain pipework. Screens can become blocked over time with scale and should be checked during 'Planned Inspections' and cleaned if necessary.

The check valve will prevent product from backing up onto the roof in the event of an internal hose failure or joint leakage when the internal roof drain valve at the base of the tank has been closed. Excessive product or rainwater on the roof can cause it to sink.

9) Visually check *pontoon compartments* for the presence of water or oil. Pontoon integrity is vital to prevent escalation in case of fire. Test for flammable gas (an indication of leaks) at least every quarter. It is good practice to make sure that each pontoon is identified by a letter or number near its' manhole on all tanks. Manhole covers on pontoons should be tightened down on their fastening lugs.

ACCIDENT This rim seal fire escalated quickly to a full surface fire when vapours contained in leaking pontoons exploded. While the rim seal fire might have been dealt with, the full surface fire proved difficult to extinguish because of a lack of water resources. The site had no routine practice of gas-testing pontoons regularly and therefore, escalation was inevitable, once undetected leaks allowed vapours build-up in the pontoons.

Early stage of the rim seal fire as captured by security camera. Note flying pontoon plate in yellow circle

Fire once the roof lost buoyancy

Example of poor design of a pontoon manhole cover: There is no gas-testing hatch. (Also note that the foam dam is lower than the secondary seal which was fitted later to the tank for environmental reasons. This denotes poor Management of Change and poor understanding of foam systems.)

Manhole covers should be secured as loose covers can float away when the roof starts to sink, or they can be blown away by wind or fire water streams. A gas-testing hatch should be provided for each compartment.

ACCIDENT This 50,000 m³ gasoline tank roof seal was changed from liquid filled type to gastight spring loaded type. This change obliged the bumpers under the roof (that hit the tank shell when the seal is fully compressed) to be modified too. It appears that these new bumpers were closer to the tank shell and made contact more often as the springs could not take the same energy as the liquid filled seal. Over the next eight years, the pontoons welds were deformed where the bumpers were attached and the roof finally sunk when three pontoons were flooded. The site had no routine practice of gas-testing pontoons regularly and therefore did not detect the pontoons flooding.

Three months were needed to safely empty and gas free the tank with minimal damage.

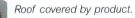

Roof covered by product.

> **Make sure a rigorous Management of Change program is in place.**
> **Visually check and gas test pontoons regularly.**

10) Check *seal, weathershield and metallic shunts* for damage and tightness to reduce the risk of fire from lightning strikes.

11) Ensure that the *lower internal roof drain valve* (located at the base of the tank) is greased, and operate it to ensure it is in working order.

12) Check *emergency roof drains* on double deck floating roofs to ensure they are free from obstruction.

13) Check the condition of *wax scrapers* to remove residues on the walls of floating roof tanks.

14) Check *guide/gauge pole rollers* for grooving to ensure that the floating roof cannot hang up and/or tilt.

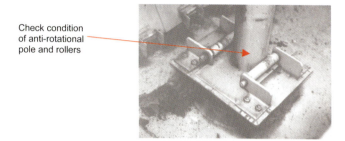

Check condition of anti-rotational pole and rollers

15) Check the condition of roller *ladders* and grease moving parts, for example, wheel and track, to ensure that they will not hang up or jam and cause the floating roof to sink.

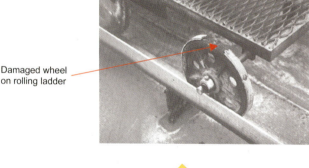

Damaged wheel on rolling ladder

Result of disconnected ladder on an open top floating roof. Failure was due to seized rolling mechanism on roof.

16) Check the condition of *foam pourers and foam dam.*

Foam pourers blocked

Special notes!

- When the floating roof is in a high position the opportunity should be taken to check the following:
  - drain sump;
  - pontoon compartments (including LEL testing);
  - seal and weathershields;
  - emergency roof drain;
  - rolling ladder;
  - metallic shunts (lightning protection).

  Such inspections should be authorized though a work permit.
- Binoculars can be used to good effect for viewing details on floating roofs.
- The tank and its roof must be in a safe condition (for example, no explosive atmosphere in the vapour space and roof not too thin for access) before such inspections are carried out.
- A standby attendant, to raise the alarm in case of an emergency, is required before operators go onto a floating roof for inspection.
- Access onto floating roofs should preferably only be authorized at high level (5 ft/1.5m) from the top. A confined space entry permit will be required if the tanks contain toxic materials, such as H_2S, or if the roof is at a lower level.

continued

- Beware of H_2S hazards from certain crude oil tanks. Breathing apparatus may need to be worn before accessing the tank's floating roof.
- Safe entry on floating roofs must be taken seriously. Refer to **API Publication 2026** for details on the general precautions and potential hazards associated with entry onto floating roofs, as well as procedures to be followed prior to entry.

A confined space entry permit and fall protection equipment are required to access the floating roof on this tank. (The roof level is more than 5 ft (1.5m) from the top.)

ACCIDENT **Roof of storage tank sinks due to no maintenance!!!**

The floating roof of a distillate tank sank during heavy rains. At the time of the incident, the tank was feeding the gasoline blender, when the gauge suddenly dropped from 26 ft to 3 ft (8m to 1m). During the hours before the roof sank, the tank's gauge stuck and remained static for approximately 11 hours.

After receiving internal complaints of odour that morning, operators began to look for the source and discovered product escaping from the tank's roof drainage system. The floating roof was found to have sunk out of sight.

The most efficient way to prevent tank fires or major leaks is to perform REGULAR visual inspection by operators.

Operating personnel should undertake the following practices:

- monitor changes in level;
- respond immediately to high level or low level alarms;
- react to any level alarms (even if trips are provided);
- on a high-high alarm, visually check tank immediately (overflowing into bund);
- prevent frequent roof landing to avoid air entry or leg damage;
- always HAZOP routine and non-routine operations;
- safeguard against product transfer errors (high RVP, hot product, etc.);
- maintain clear and up-to-date emergency operating procedures and provide refresher training/drills for operators.

7.6 In-service inspections

In-service inspections are undertaken for external plate thickness measurements to determine the rate of corrosion. All tanks shall be given a formal external inspection by a qualified inspector at least every five years. Tanks may be in operation during this inspection.

For open top floating roof tanks, external examinations are undertaken when the roof is near the top of the tank to ensure good ventilation since when the roof is at low level, natural ventilation is poor and breathing apparatus must be worn to access the roof.

Insulated tanks need to have the insulation removed only to the extent necessary to determine the condition of the exterior wall of the tank or roof.

If there is any reason to believe that the roof is weak and could fail under a man's weight, full harness and fall arresting equipment must be worn when carrying out inspections.

Ultrasonic thickness measurements of shell and roof

External, ultrasonic thickness measurements of the shell is a means of determining a rate of uniform general corrosion while the tank is in service, and can provide an indication of the integrity of the shell.

The ultrasonic thickness measurements shall be performed at intervals not exceeding the following:

- five years after commissioning new tanks;
- at five year intervals for existing tanks where the corrosion rate is not known or in accordance with a Risk Based Inspection (RBI) schedule.

Remote operated, ultrasonic crawler allows cost effective thickness measurements on carbon steel above ground storage tanks without the need for costly scaffolding or rope access services. The low weight crawler is coupled with magnets to ensure good adhesion to both the wet and dry surfaces— vertically, horizontally and even inverted without the loss of inspection capabilities. The use of lightweight cable allows inspection at heights of 100 ft (30m) or more. Exposed tank roofs can be inspected without the risk of falling.

 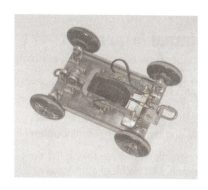

Crawler that can climb a tank and measure the tank shell wall thickness.

ACCIDENT **Fatal fall through a corroded roof during inspection!!!**

Two contractors climbed on a cone roof tank that was to be fitted with a new roof to evaluate the extent of the work. The roof was heavily corroded and the tank had been emptied, gas freed and cleaned. Despite clear warning signs posted on all access ways, they walked on the roof and one of them fell through the corroded plates to his death, 15m (49 ft) below.

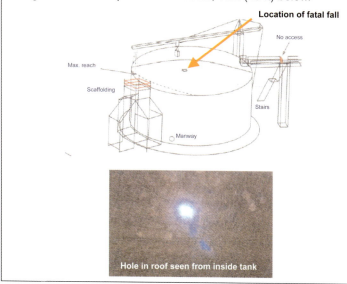

125

Acoustic emission testing is now becoming a recognized technique to monitor the condition of the tank floor/bottom. However, it requires specific conditions (over 50% full, time to settle, no wind or rain, no background noise) and specialist technicians to carry it out. It is a qualitative method that can prioritize tanks for internal inspections or provide justification for keeping tanks in service for additional years. The bottom/floor plate can be assessed without the need to take the tank out of service.

7.7 Internal inspections

Internal inspections are carried out primarily to:

- ensure that the tank bottom is not severely corroded or leaking;
- gather the data necessary for the minimum bottom and shell thickness assessments;
- identify and evaluate any tank bottom settlement.

Internal inspections require taking the tank out of service and this can be done in compliance with national standards or an RBI assessment.

Intervals between internal inspections shall be determined by the corrosion rates measured during previous inspections or anticipated based on experience with tanks in similar service. Normally, bottom corrosion rates will control and the inspection interval will be governed by the measured or anticipated corrosion rates and the calculations for minimum required thickness of tank bottoms.

In no case, however, shall the internal inspection interval exceed 20 years. When corrosion rates are not known and similar service experience is not available to determine the bottom plate thickness at the next inspection, the inspection shall be carried out within ten years.

It is good practice to internally inspect tanks as follows:

- in corrosive service at least every five years;
- containing white products (e.g. gasoline, jet fuel, gasoil) every ten years;
- containing black products between 10 to 20 years depending on water content.

> **Maximum internal inspection is 20 years if corrosion rate is known and not found to be critical, or ten years if corrosion rate is unknown.**

A magnetic flux scanner can be used to determine the state of the floor/bottom plate. It provides a colour-coded display of corrosion patterns as shown below. The figure on the right represents the floor condition with each 250×250mm (9.8 in $\times$ 9.8 in) square being colour-coded for the worst defect in the plate. It is used to quickly identify which plates have defect indications above the minimum reporting level. The legend shows the colour-code representing a percentage material loss of the nominal floor plate thickness. It is also used for marking up areas to be repaired.

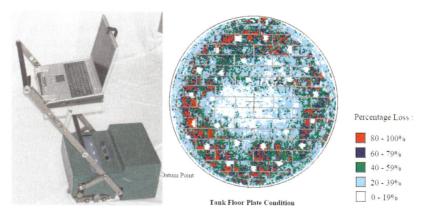

Datum Point

Tank Floor Plate Condition

Percentage Loss :

- 80 - 100%
- 60 - 79%
- 40 - 59%
- 20 - 39%
- 0 - 19%

Left: tank bottom thickness measurement.
Right: the measurement in graphic output.

ACCIDENT Corrosion on floor plates results in tank leak!!!

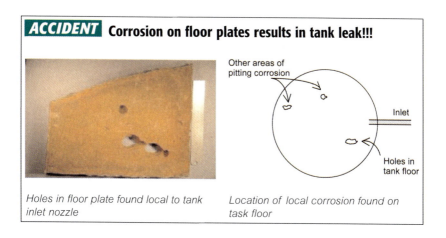

Other areas of pitting corrosion

Inlet

Holes in tank floor

Holes in floor plate found local to tank inlet nozzle

Location of local corrosion found on task floor

The bottom floor should be sloping towards the sump in order that all water is drained off regularly. During internal inspections, after floor plate repairs, this should be checked by the 'Inspection Authority' to ensure no possibility of water pooling.

127

ACCIDENT Spent sulphuric acid tank failure!!!

An explosion at a refinery killed one maintenance contractor and injured eight others. The contractors were repairing the grating on a catwalk in a sulphuric acid storage tank farm when a spark from their hot work ignited the flammable vapour in a spent H_2SO_4 tank. The tank exploded with the roof and shell separating from its floor, instantaneously releasing its entire content of H_2SO_4. Other tanks in the common bund were affected and also released their contents. A fire broke out and lasted for half an hour. The acid spill reached a nearby river resulting in significant damage to aquatic life.

The maintenance of the tank was inadequate. It had many holes in the roof and shell and had never received an internal inspection. The holes in the tank allowed ignition sources to come into contact with flammable vapours.

API 653 requires a ten-year internal inspection interval or earlier depending on the corrosivity of the stored material. The refinery inspectors recommended an earlier inspection but this never occurred. Implementation of the tank inspection program was inadequate. NACE (National Association of Corrosion Engineers) recommends internal inspection every five years for tanks in concentrated sulphuric acid service.

Spent acid tank explosion separating roof and shell from the bottom plate

continued

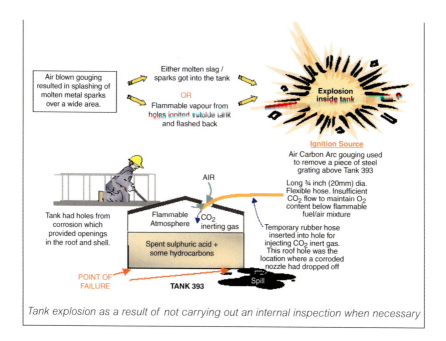

Air blown gouging resulted in splashing of molten metal sparks over a wide area.

Either molten slag / sparks got into the tank

OR

Flammable vapour from holes ignited outside tank and flashed back

Explosion inside tank

Ignition Source

Air Carbon Arc gouging used to remove a piece of steel grating above Tank 393

Long ¾ inch (20mm) dia. Flexible hose. Insufficient CO_2 flow to maintain O_2 content below flammable fuel/air mixture

AIR

Flammable Atmosphere

CO_2 inerting gas

Temporary rubber hose inserted into hole for injecting CO_2 inert gas. This roof hole was the location where a corroded nozzle had dropped off

Tank had holes from corrosion which provided openings in the roof and shell.

Spent sulphuric acid + some hydrocarbons

POINT OF FAILURE

TANK 393

Spill

Tank explosion as a result of not carrying out an internal inspection when necessary

129

8

Taking tanks out of service for maintenance and other purposes

The purpose of taking tanks out of service includes:

- mandatory internal inspection;
- maintenance and repair;
- cleaning;
- change of products.

All tanks that have contained chemicals or petroleum liquids may present one or more of the following hazards during some phase of preparing a tank for cleaning, maintenance or during recommissioning:

- fires and explosions;
- oxygen deficiency or enrichment;
- toxic substances (liquids, vapours, fumes and dust);
- physical or other hazards.

8.1 Planning is the KEY

Each employer should develop and implement appropriate administration controls and written plans for tank work from decommissioning through recommissioning that address the following:

- Work plan (Refer to API 2015 and 2016 for details and additional guidance).
- Pre-job discussions with representatives of involved parties and contractors before the work commences.
- Use of qualified persons at the job site to supervise and perform activities.
- Training of personnel in precautions and procedures.
- Use of adequate safeguards to control hazards for protecting personnel and equipment.
- Use of permits, such as cold work, confined space entry and hot work.
- The use of degassing equipment to prevent or reduce the amount of organic volatile compounds/hazardous material released into the atmosphere during vapour/gas freeing, for example, activated carbon and thermal oxidation units.
- Use of appropriate ventilation procedures.

- Testing by a qualified person of the tank's atmosphere before and during entry periods.
- Emergency and Rescue Plan.
- Compliance with applicable company policy, procedures, local legislation and API 2220 *Improving Owner and Contractor Safety Performance*.

8.2 Storage tank plan elements

Before a tank is taken out of service, the responsible supervisor should prepare a plan of the various steps/stages involved in the process. This plan should include a risk assessment (Job Hazard Analysis of all tasks) and a checklist. Each step should be controlled through the WORK PERMIT SYSTEM.

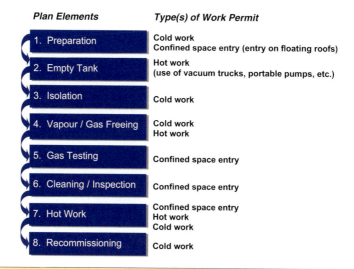

Plan Elements	Type(s) of Work Permit
1. Preparation	Cold work Confined space entry (entry on floating roofs)
2. Empty Tank	Hot work (use of vacuum trucks, portable pumps, etc.)
3. Isolation	Cold work
4. Vapour / Gas Freeing	Cold work Hot work
5. Gas Testing	Confined space entry
6. Cleaning / Inspection	Confined space entry
7. Hot Work	Confined space entry Hot work Cold work
8. Recommissioning	Cold work

Removal of a storage tank from service to recommissioning may take many months. Each step in the process must be authorized through the work permit system and the various tasks subject to a Job Hazard Analysis to assure the adequacy of the proposed safeguards.

Stage 1: Preparation

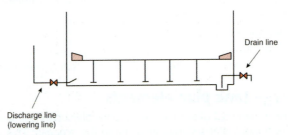

Obtain engineering and piping drawings. These drawings will show in detail the internals of the tank that will determine the method for pumping out, draining and isolation.

Undertake all safety requirements that must be carried out while the tank is in service. For example:

- In the case of open top floating roof tanks, the roof legs should be inspected when the tank is at high level to ensure that they are in a safe working condition and then lowered to their maintenance position;

- All bolts and nuts should be removed and replaced *one by one* at all manholes, mixers, pipework connections etc., so that they can be removed easily at the time when the tank is empty.

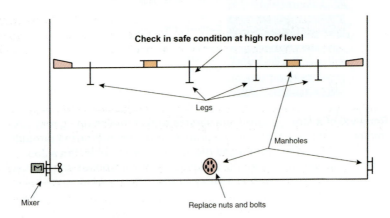

Stage 2: Empty tank

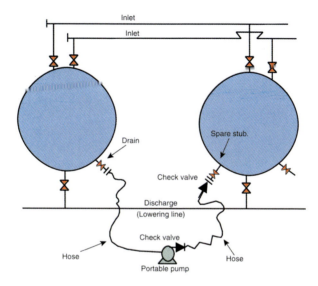

Tanks are initially emptied by means of the normal lowering outlet until suction is lost. When the drain lines are connected to a closed system, such as a sump tank, the remainder of the tank contents can be drained.

When closed system drains are not fitted, a flexible hose attached to the drain via a pump can be used to transfer the contents either into the normal lowering line or to a suitable alternative, for example, a road tanker truck, adjacent tank (see the diagram above).

Make sure that:

- positive displacement pumps are protected by discharge relief valves;
- flexible hoses have flanged ends and are suitable for the service;
- non-return valves are in the right way round.

When a portable/mobile pump driven by a motor or diesel engine is used for emptying a tank, a *hot work permit* must be obtained specifying a safe location for the pump. These types of pumps can produce a spark causing a fire or explosion. If the product is flammable, the following types of pump should only be used:

- air driven pump;
- steam driven pump;
- hydraulically driven pump.

At this stage, the entire product should have been removed from the tank leaving a layer of water below a layer of product vapour.

Stage 3: Isolation

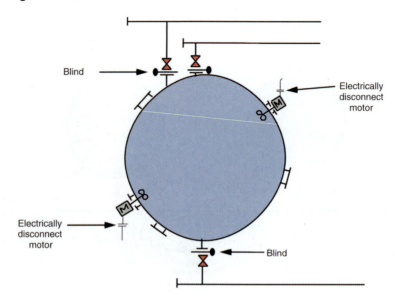

Important points to note include:

- Blind off and electrically isolate (lockout/tagout) the tank.
- Blinds should be placed on the tank side of the isolation valve to prevent any possibility of product entering.
- Wear correct personal protective equipment.
- Wear fire resistant coveralls.
- To increase ventilation and improve rescue capabilities, it is preferable to remove side mixers and disconnect sections of large bore piping, blanking off the open ends.

Stage 4: Vapour/gas freeing

Vapours can either be discharged to atmosphere at a high level or to an approved degassing system (thermal oxidation, refrigeration or activated carbon) dependent upon local environmental regulations.

Tank degassing using activated carbon

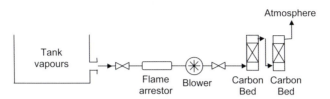

For direct discharge to the atmosphere, mechanical ventilation devices (air movers, blowers, etc.) must be fitted to manways and the exhaust outlet elevated to ensure adequate dispersion of the vapour. Install air blowers and exhaust outlet during final water draining to minimize loss of vapour to atmosphere while fitting these devices.

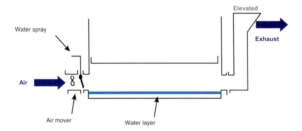

During the vapour-freeing process, the time during which the vapour/air mixture in the tank is in the flammable range should be minimized. For light products such as gasoline, the vapour inside the tank after gas freeing (with the air blowers shutdown for approximately 15 minutes) should be as follows at this stage:

- LFL/LEL—less than 1%;
- Oxygen content—21% (similar to air).

Determine whether there is a possibility of pyrophoric deposits in the tank. These deposits should not be allowed to dry out because they may spontaneously generate heat and ignite causing explosion and fire. All interior surfaces can be kept wet by the use of water sprays. See sketch above. Refer to BP Process Safety Booklet *Safe Ups and Downs for Process Units* for more details on this subject.

> **Beware of pyrophorics!**
> **They may generate heat and ignite causing explosion and fire.**
> **It is important to keep any pyrophoric scale WET until the material can be removed to a safe location.**

Stage 5: Gas testing

Inadequate or improper gas testing can be a cause of fatal accidents because many people do not understand the basic working principle of flammable gas detectors. Most portable flammable gas detectors operate by the catalytic combustion of a flammable gas on a heated filament (usually platinum), to give a reading of the %LEL (or %LFL). Accordingly, there must be approximately 21% oxygen in the sample to give an accurate reading. If the atmosphere being tested is deficient in oxygen, for example, purged with nitrogen, it is not possible to use a standard type of flammable gas detector to detect hydrocarbon vapours.

Flammable gas detectors must be maintained, calibrated and operated in accordance with the manufacturer's recommendations in order to give accurate results.

Not for %LFL testing!

Always use the correct instrument for the measurement of oxygen, LEL and percentage of toxics. Do not use gas tube detectors (which change colour upon detection) for measurement of %LFL. This device is used to determine the presence of low concentrations of a known contaminant as each tube is designed for a specific contaminant.

Similarly combustible gas detectors should not be used to measure toxic substances as they will not measure the very low concentrations of toxic substances present (in parts per million) that may pose a health risk.

Notes

- Gas testing must be undertaken in the following order:

- If the gas testing equipment is calibrated on one specific flammable mixture, it may be inaccurate for measuring the LEL for another mixture, for example, if calibrated with methane/air mixture of 10% LEL it may give a reading of 29% LEL for toluene/air mixture.
- A low O_2 reading means the presence of a contaminant—find out what it is. Something is wrong. If the reading is 19.5% O_2, then the atmosphere contains 6.25% of another contaminant.
- If O_2 is above 20.8%, which is the normal O_2 content in air, there is a leak from an oxygen cylinder inside the tank.
- Make sure all areas of the storage tank/vessel are gas tested and visually inspected. Refer to Stage 7.
- Remember that hydrocarbon vapours are heavier than air and tend to settle in low places.

*Conditions for entry without respiratory protection:**

- The confined space is clear of all deposits, scale and sludges likely to give off vapours when disturbed.
- The work to be done inside the confined space will not release or generate flammable or toxic vapours.
- The confined space is adequately ventilated and tested to ensure a safe atmosphere.

*Note: Air purifying respirators may only be worn in a confined space under specified conditions defined by the safety/industrial hygiene specialist or respiratory program co-coordinator.

Conditions for entry with air supplied respiratory protection

- Potentially harmful deposits, scale or sludges are present which cannot be removed by outside cleaning.
- All practicable measures have been taken to eliminate flammable or toxic vapours.
- Ventilation is maintained to reduce the concentration of contaminants to a minimum.
- Constant monitoring of the confined space is required to ensure harmful contaminants remain within acceptable limits.
- Authorized entrants must be withdrawn immediately if these conditions are exceeded and the conditions must be re-evaluated by the Issuing Authority before any subsequent entry.

> **A Risk Assessment is required prior to entry into a tank for gas testing, and controlled through a Confined Space Entry Permit in accordance with the site's Safety Rules.**
>
> **Gas tests and a visual inspection will be required to determine whether there is a safe working environment for hot work.**

A standby attendant must remain on duty outside the tank for the duration of the entry. Standby attendants should be given proper training and provided with:

- Full instructions on their duties.
- Appropriate PPE and breathing apparatus—so that they can look into the confined space.
- Strict instructions not to enter the tank under any circumstances. Their role is to raise the alarm and wait for rescuers to arrive and then communicate vital information to the rescue team.
- A copy of and training in the emergency/rescue plan.

> **Standby attendants are strictly not allowed to enter the tank under any circumstances!**

Refer to the BP Process Safety Booklet *Confined Space Entry* for more details on this subject.

Diagram showing authorization requirements for entry against various levels of gas concentration

	Oxygen	Flammable	Toxic
Entry without breathing apparatus (BA)	20.8%	<1% LEL	<10% OEL
Special risk assessment for use of BA	19.5–23.5%	Up to 20% LEL*	Up to STEL
Unsafe—No entry allowed	>23.5%, or <19.5%	>20% LEL	> STEL

* Flammable gas concentration should be below 10% LEL (this should be considered as an entry criteria, with 20% LEL used as an exit criteria where work should stop and persons be withdrawn from the confined space).

LFL—abbreviation for 'Lower Flammable Limit' which is the lowest concentration of a flammable gas in air capable of being ignited by a spark or flame. The term 'Lower Explosive Limit', or 'LEL' is sometimes used to describe the same effect.

OEL—abbreviation used for Occupational Exposure Limit value used to restrict total long term intake by inhalation over one or more work periods to levels to which an average worker may be exposed without harmful effect. This is normally quoted as the time weighted average (TWA) concentration over an eight-hour working day. National Authorities publish their own exposure limit type data (for example, the UK Health & Safety Executive publishes Occupational Exposure Limits, and in the US TLV's (threshold limit values) are published by ACGIH).

STEL—abbreviation used for Short Term Exposure (usually not exceeding 15 minutes) which is applied to restrict brief exposures over the OEL which are unlikely to produce serious short or long term effects on health over the time it is reasonable to expect the excessive exposure to be identified and remedied. For substances assigned both an OEL and STEL the total number of exposures above the OEL should not exceed one hour in a 24 hour period (i.e. four per day), and should not cause the OEL measured as a time weighted average to be exceeded. Again, National Authorities publish their own exposure data as indicated above, and describe methods of determining short term exposure limits if none are quoted in the standard.

Life Threatening Atmospheres—no person should be allowed to enter a life threatening atmosphere with, or without, normally available respiratory protection as this does not provide adequate protection to work safely. A life threatening atmosphere can be considered to be one where:

- the oxygen content is below 19.5% or above 23.5%;
- concentrations of flammable gas or vapour are above 20% LEL;
- concentrations of toxic gases or vapours are above short term exposure values (STEL).

Where national standards are more stringent, these must be followed.

Stage 6: Cleaning

Specialist tank cleaning contractors should only be used.

Due to the potential high risk, the contractor should be required to submit a written site safety plan describing exactly how the work will be done including:

- confined space entry procedures and permit requirements;
- method of treating sludge and scale;
- site layout drawing showing equipment placement such as the drawing shown below.

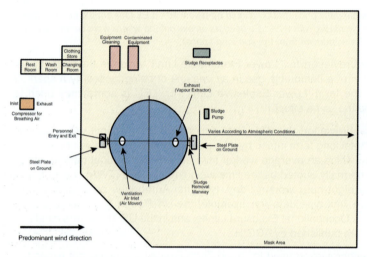

An important safeguard during tank cleaning operations involving flammable residues is the provision of adequate ventilation. This should be sufficient to reduce the overall flammable vapour concentration, avoid vapour pockets, and minimize the vapour concentration above the sludge surfaces.

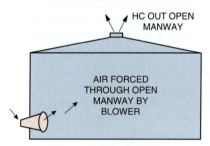

A method of venting cone roof tanks is to place a blower in the shell manway and force air into the tank allowing the vapour air mixture to escape through the roof manway.
The blower or eductor should be bonded to the tank to prevent static build-up

As much sludge and residue should be removed from the tank without entering it. However this may not always be practicable so at sometime it will be necessary to remove the remaining residue from within.

The precautions that need to be taken shall be stipulated by the Confined Space Entry Permit Issuing Authority and subject to a Job Hazard Analysis.

These precautions will include but not be limited to:

- type of respiratory protection to be worn;
- entry requirements (names of entrants, standby attendants and rescuers);
- continuous ventilation;
- continuous atmospheric monitoring;
- prevention against heat stress;
- specific type of equipment to be used.

ACCIDENT **Smoking in confined space results in fatality!!!**

A 78m (286-ft) diameter floating roof tank used for crude oil storage had been taken out of commission for cleaning, inspection and repair. An Entry Permit and a Hot Work Permit (for diesel engines etc. located in the bund area) had been issued.

Three employees of a contracted specialized tank cleaning firm were working inside the tank removing crude oil sludge, wearing air-line breathing apparatus, when a fire started inside the tank. Two of the contractor's employees escaped from the tank immediately but the third man failed to exit.

The fire was extinguished by the Safety Services operator, assisted by Terminal staff before the Fire Brigade arrived some twelve minutes later. The body of the trapped person was subsequently recovered.

It appeared that the contractor's employees had started smoking inside the crude oil tank, which ignited the residue material in the tank. An inquiry confirmed that the contractor's employees on site knew that smoking was only permitted in the designated areas but had broken this rule.

Special Note: If working on tanks requires special attention to the hazardous atmosphere they may contain as described above (hot work, smoking, etc.), one must not forget the more conventional threats to the workers such as lifting, working at height, etc as in the following three accidents.

ACCIDENT A supervisor wanted to check the progress of work on a floating tank, which had been taken out of service. He climbed up the stairway to check the roof. Unseen by the standby attendant, who continued up the stairs to the top of the tank, the supervisor climbed over the handrail onto the wind girder of the tank. He lost his balance and fell approximately 9–10m (30–33 ft) from the wind girder to the ground.

Coled Air Hoses
Saved Fall

Amazingly, he fell onto a pile of coiled air hoses, which were lying on the sloped section of the tank base (see photograph). He miraculously survived the fall with only fractured ribs, shoulder blade and hip, which did not require surgery. He was released from the hospital after two weeks.

ACCIDENT During tank construction, a scaffolder fell 18m (59 ft) to his death as he was transferring from the outer ring of the staging to the inner one over the top of the shell plate to reinstate the scaffold to permit welding of the roof sections to the shell plate.

The top of the shell plate was approximately 1.05m (41 in) above the staging boards. As he landed on the middle scaffold board/plank of the inner staging, it snapped along a line of knots. The other two boards/planks separated as they were not fixed into position allowing a gap large enough for him to fall through to the tank floor below.

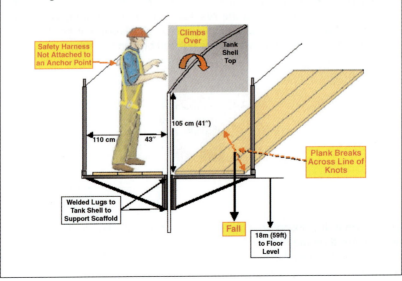

ACCIDENT A contractor was performing a diesel wash on top of a tank internal floating roof (aluminum honeycomb) in preparation for API 653 inspection. The tank had been taken out of service for cleaning and inspection. The roof was sitting on its high legs. A technician was located on the south side of the internal floating roof when the north end of the roof split and partially collapsed. The technicians immediately evacuated the tank.

The high position support cap did not have full thread engagement with the tank's internal roof support bracket. Therefore, the roof failed to the lower leg support. All other internal roof legs were in the secured high leg position.

Key messages:

- ensure high poisition support caps have full thread engagement;
- provide temporary supports under roof prior to roof cleaning and inspection.

Stage 7: Hot work

After the tank has been cleaned, it must be inspected and gas tested again in preparation for repairs/hot work.

*Examples of potential hazards that can remain **after** cleaning*

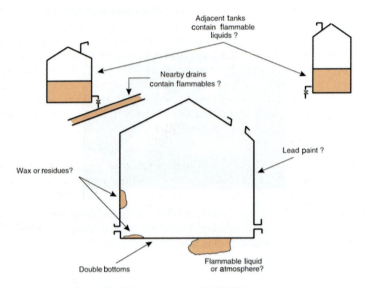

Have you checked for these potential hazards after cleaning your tank?

Gas tests and internal inspections must be carried out to ensure that there are no hidden hazards! A rigorous check of all areas where hydrocarbons may remain must be carried out before relaxing entry requirements with respect to breathing apparatus and allowing hot work. If there are holes under the floor, hydrocarbons will have accumulated and will migrate back into the tank.

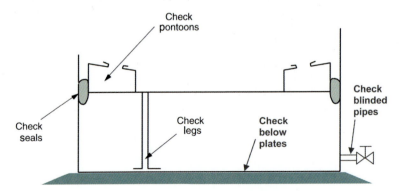

Check pontoons, seals and legs for flammable atmosphere during gas testing.

ACCIDENT Explosion on floating roof tank during hot work!!!

An explosion occurred during repair work on the wall of a naphtha floating roof tank. This resulted in a 30m (98 ft) stretch of both the upper and lower parts of the double seal and tank roof being damaged.

Eleven years ago, the single seal was replaced by a double seal—this modification required a change to the construction of the roof rim, resulting in the creation of a hollow space below the new seal, where hydrocarbon could accumulate. This space was not self draining and was difficult to reach with normal cleaning equipment. Prior to the repair work, the tank had been taken out of service, emptied, cleaned, inspected, gas tested and refilled with water to raise the floating roof to the level required for execution of the repair. Although gas tests were performed daily inside the tank, on top of the roof, and around the circumference, the space in between the seals and underneath was not tested.

The repair entailed replacement of a section of the tank wall requiring cutting, grinding and welding. A hot work permit had been issued prescribing covering the roof seal with fire blankets. However, these blankets had not been fixed properly, leaving part of the seal exposed. There was local damage of the seal indicating that sparks had fallen on the seal and possibly had found a way to the space below.

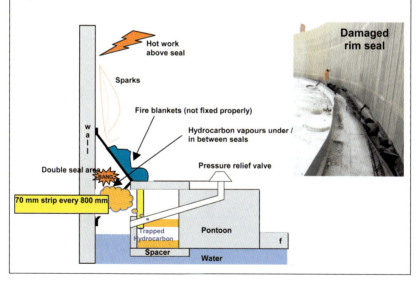

Points to note during hot work:

- Possibility of flammable liquids/vapours in the ground beneath the tank floor and in double bottoms—ensure that there is no flammable or other mixture under the tank floor.

- Cleanliness of all parts of the tank—waxes and other deposits that will emit vapour must be removed.

- Contents of nearby storage tanks, drains and sewers (fire risk).

- Toxic fumes from welding/cutting operations—maintain effective ventilation such as using local exhaust ventilation to remove welding fumes.

- Leave all gas cylinders and welding machines outside the tank.

- Ensure lighting is adequate with ground fault circuit interruptors.

- Make sure contractors do not create flammable/toxic atmospheres during painting.

- Continuous gas testing in and around the tank.

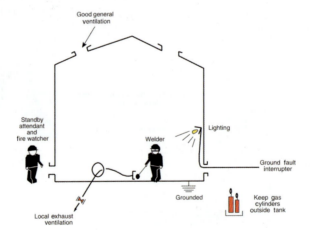

ACCIDENT **Argon arc welding in confined space causes oxygen deficiency!!!**

A welder suffered headaches after a day inside a tank using argon to weld repair titanium. It was discovered that there was a rapid decrease of oxygen in the area where he was working due to a poor ventilation arrangement. Furthermore, a manway watch attendant was monitoring the oxygen content in the tank at all times, but the 10 ft (3m) wand he was using was not long enough to monitor the air where the welder was working (see figure below).

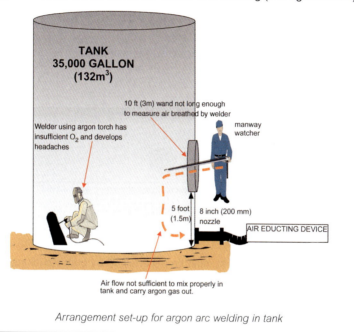

Arrangement set-up for argon arc welding in tank

Local exhaust ventilation is required for welding work in a confined space.

The minimum safe level of oxygen in the air for working is 19.5% (which is not far below the normal level in the air of 21%). Refer to BP Process Safety Booklet _Hazards of Nitrogen and Catalyst Handling_ for more details on the dangers of inert gas atmospheres and working in confined spaces.

ACCIDENT Fire in floating roof tank under maintenance!!!

As maintenance personnel were in the process of cutting an access door in the wall of a tank, a fire erupted.

A maintenance crew had separated the primary and secondary seals from the tank wall to avoid possible heat damage when the wall was cut. An accumulation of oily liquid was released and not discovered.

No further internal inspection was carried out after the two seals were separated leaving a layer of oil inside.

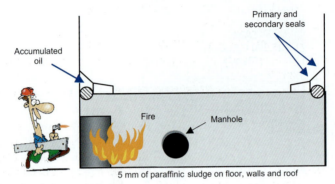

Fire in tank under maintenance

Lessons learned

- Cleanliness of equipment must be ascertained by both visual inspection and gas testing before issue of hot work permits.

- Care must be taken to ensure that 'trapped' pockets of oil, sludge, scale, which cannot be determined by gas testing alone, are not vaporized by hot work to give a flammable mixture with air leading to fires/explosions.

- Strict observance of well-established safe procedures for cleaning equipment is essential, paying extra attention to recesses, behind linings, and any other trapped area.

- Special techniques may well have to be applied to clean and gas free such voids before attempting adjacent hot work.

- The safety of personnel, with particular reference to means of escape must be considered when performing hot work in or on tanks. It is essential that any hot work is carefully planned and safely integrated with other work to be done.

Stage 8: Recommissioning

- Plant inspectors must check the integrity of any repairs in accordance with API 653.
- This may require a hydrotest and therefore a requirement for a plan to ensure the correct purity of water, safe filling/discharging rates and proper disposal of the effluent.
- A thorough inspection must be taken by operations and other departments to ensure the correct reinstatement of:
  - ○ electrical equipment (grounding/earthing, cathodic protection);
  - ○ instrumentation—high level alarms and tank gauging;
  - ○ fire protection—live-tested;
  - ○ mechanical—all components properly reinstated;
  - ○ operations—all valves are in their correct positions and correct mesh sizes on vents.
- If there is to be either a full or partial water test of the tank, it should remain isolated from the process pipework system.

ACCIDENT An accident occurred when the roof of a Jet A1 tank began to be covered with product after heavy rain. The tank had just been put back in service after routine inspection and repairs. These repairs included changing some metal sheets on the single deck roof. The investigation revealed that during that job, the contractor replaced the emergency drain pipe (which is supposed to send rain water into the tank in case the normal drain is closed or plugged, to prevent overloading of the roof) by a longer pipe than the original one. Therefore, more rain was allowed to stay on the roof. As a result, the weight of rainwater forced Jet A1 to flow back through the emergency drain and flooded the roof.

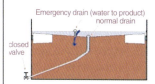

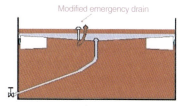

Normal design of the emergency drain: In case of heavy rain, water is allowed into the tank to prevent sinking the roof.

Emergency drain modified: Weight of rainwater pushes the roof down, forcing product on top of the roof.

However small and cheap a modification appears to be, it must still be subjected to your site's 'Management of Change' procedure to ensure all potential hazards have been adequately addressed.

Special Note: Emergency drains on single deck floating roofs are *not* recommended because of the risk of product backflow onto the roof. These tanks should be equipped with enough sufficiently sized normal rainwater drains with outlet valves kept opened, and regularly inspected to ensure their continued integrity.

ACCIDENT A MTBE cone roof tank with internal floater was taken out of service for routine repair and maintenance. One month later, this routine work was completed, at which point the tank was then inspected and signed out ready to be put back into service. All the blinds were swung and the tank side rundown valve was also opened in preparation for recommissioning of the tank.

MTBE supply delays meant that the tank would not be required at that time and this was communicated to operations. About ten days later, a severe electrical thunderstorm developed over the area and there was a direct lightning strike to the tank, immediately followed by an explosion which caused the tank to jump up in the air and the fixed roof to shear off (see photo).

Investigations found that several gate valves (not tight shut off-valves) around the tank were leaking. With all the blinds removed, enough MTBE would have leaked into the tank to create an explosive mixture. In view of the explosive range of MTBE in air, an amount of about 40 litres of MTBE would have been enough to reach the lower explosion limit for the space under the floating roof. It was concluded that a small amount of MTBE leaked into the tank and evaporated over a ten-day period prior to a direct lightening strike that ignited the explosive mixture.

Lessons learned

- Isolation blinds should only be turned just prior to recommissioning when product is available to reduce the exposure time in which a flammable mixture could be present.
- Avoid decommissioning or recommissioning tank activities for volatile mixtures during periods of electrical storms.
- Barricade or restrict access to the area during the high-risk periods of decommissioning and recommissioning tanks.

Other safety and environmental matters

9.1 Oil spills–prevention and response

Any location where a spillage can occur must be equipped with a containment and recovery system to prevent contamination of the environment.

These locations include:

- jetties where hose/loading arm connections are made;
- road/rail truck loading rack;
- road/rail truck discharge point;
- drum storage and handling area;
- pumps and manifolds;
- bulk storage tanks;
- drain and sample points.

Such collection/containment systems will need to connect to a treatment system which may include one or more of the following:

- holding tank;
- settlement pond;
- oil and water separator;
- parallel plate interceptor separator;
- corrugated plate interceptor separator.

It is essential that these treatment systems are designed and maintained to eliminate any potential discharge of oils, greases or chemicals to waterways or sewers. This includes the capacity to handle the firewater run-off that may be used in a fire. The quantity of water expected to be applied should be provided in the pre-fire plan.

Good practice

Proper wastewater handling at a treatment system for handling of oily water collected in sumps and drains.

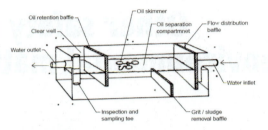

American Petroleum Institute (API) Type Separator

An API separator is a piece of equipment that utilizes the difference in density between oil or petroleum products and water to remove the oil or associated chemicals from the water. It is used to reduce the amount of effluent water contaminated with oil or other petroleum products releasing into the environment and to purify water for reuse, as part of environmental protection efforts. The oil floats to the surface of the water, where it can be skimmed off for recovery.

However, some other hazards associated with API separators or other water treatment facilities are:

- drowning hazard (open pits);
- fire and toxic gas exposure, particularly H_2S;
- high corrosion of walkways and equipment.

Good example of closed API separator with carbon canister absorbing system to prevent VOC emissions to atmosphere
(Refer to Section 9.7 for the hazards of carbon canisters).

Good example of API separator with skimmer tilt controls extended out of the pit so that operators do not have to go into the pit. It also has good signage, and gas detection with a visual alarm.

Poor practices

Missing sections of retention wall to prevent spill from hose entering the sea.

Wooden jetty with gaps between planks—no retention against spills

Earth pit used to collect oily water for recovery—possible seepage into the ground

Open water treatment pits present the hazards of gassing and drowning.
Note: No life jackets available.

Wastewater treatment plants must be fitted with adequate emergency response equipment (detection, portable and trolley fire extinguishers, foam systems, safety showers, life buoys, eye-wash equipment if chemicals are handled, etc.).

ACCIDENT Fire in sewer!!!

A fire broke out in a water treatment facility when sparks from hot work fell in a nearby sewer. The fire propagated quickly to the wastewater treatment basins and tanks and caused heavy damage. Explosions occurred in the sewer system in multiple areas because of poor maintenance of the plant sewer system (note flames coming out of two different sewer covers).

9.2 Oil spill contingency plans

Every terminal must have a plan that identifies the personnel, equipment and materials it needs to deal with spills.

It should include information about environmental sensitive areas, personnel training, practice drills and a 'worst case' scenario.

Most areas of the world are covered by a mutual aid scheme with government and industry working together to respond to spills.

There are number of cleaning tools available. They include:

- booms (containment, collection and deflection);
- skimmers;
- sorbents (pads, pillows and booms);
- chemical dispersants;
- bioremediation.

> **Personnel must regularly drill in the use of the clean-up tools to ensure a speedy response in the event of an oil spill.**

One of the aims of this booklet is communicate good practice and safeguards in order to prevent leaks and spills so that the need for such emergency response never occurs.

Refer to IPIECA (International Petroleum Industry Environmental Conservation Association)'s *Guide to Contingency Planning for Oil Spills on Water* for options available to minimize environmental damage from oil spills.

Boom in action

Some clean-up tools for oil spills

9.3 Ship cargo pumproom sea chest valves

The risk of pollution through cargo shipside valves should be considered by the terminal and ship representatives when completing the ship/shore safety checklist. Refer to Chapter 5 for an incident where failure of a ship to close the sea chest valves caused spillage of 3000 tons of gasoline into the sea. A passing fishing vessel ignited the spill.

9.4 Volatile Organic Compounds (VOCs)

VOCs are defined by the United Nations as all organic compounds of anthropogenic nature, other than methane, that are capable of producing photochemical oxidants (ozone) by reactions with nitrogen oxides in the presence of sunlight. Ozone at ground level under certain exposure conditions may adversely affect human health (respiratory problems), interfere with plant growth and damage building materials. Additionally, some VOCs contribute to stratospheric ozone depletion, some may be toxic or potentially carcinogenic (for example, benzene) or pose other health risks, and most have an indirect contribution to global warming (through ozone production).

As a result, many countries have introduced legislation mandating the control of new and existing VOC emissions. With the implementation of environmental management systems (ISO14001), other facilities have taken initiatives to minimize emissions of hydrocarbon vapours from storage tanks and other sources like drains/separators as part of continual improvement. This includes the use of double rim seals, improved gasket and covers on roof fittings and the installation of geodesic covers to open top floating roof tanks to reduce evaporative losses from the effects of wind.

> **Protect the environment by minimizing hydrocarbon liquid evaporation losses to the atmosphere, for example, by keeping equipment in good order, repairing leaks immediately, keeping sample/dip hatches closed.**

Measures to minimize air emissions have also included closed venting systems and the installation of carbon canisters on vents to atmosphere to absorb hydrocarbon vapours. In addition small leaks, known as fugitive emissions, due to leaking flanges, pump glands and other connections are identified and repaired promptly.

Sampling hatch is left open—poor practice

> **Check flanges, valve glands and packings for fugitive emissions.**

9.5 Vapour balancing/return

Another method of emission control is the application of a vapour return line (sometimes called vapour balancing line) during loading and unloading operations. It is a practice involving the loading of a container where the vapour that is displaced by the incoming liquid is returned to the source of transfer through a connection between the container vent and the source tank. The vapour is 'returned' to the source tank in a closed system, hence eliminating any vapour emission to atmosphere. This practice improves personnel safety, property and environmental protection, but the systems must be carefully designed, maintained and operated.

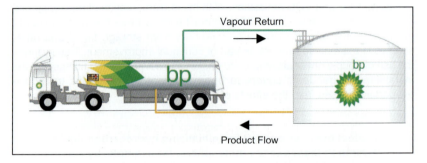

A schematic diagram of a vapour return system for truck loading

Some organic compounds are carcinogenic, such as benzene. They are known or suspected to cause cancer in animals or humans. Other organic compounds contribute to the problem of air pollution as they can participate in complex photochemical reactions to form ozone and smog.

9.6 Vapour control systems

These systems are intended to collect vapours from transfer operations to meet environmental VOC emission regulations. Vapours given off during ship loading, for example, may be recovered or destroyed. The design and installation must minimize the risk of overfill, overpressure, vacuum and pipeline detonations and deflagrations. Because of the risk of ignition of potential flammable mixtures within these types of facilities, approved flame/detonation arrestors are normally installed in the piping network close to ship as well as close to any potential ignition sources. Refer to US Coast Guard Standard (33CFR Part 154, Subpart E) or CCPS Guidelines *Engineering Design for Process Safety* for additional information on this subject.

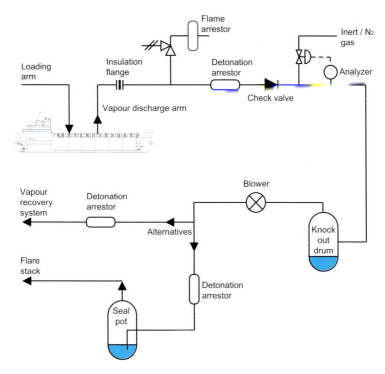

Brief outline of a marine vapour control system

9.7 Carbon canisters

Activated carbon canisters are sometimes installed on tank venting systems to absorb volatile, toxic or odorous hydrocarbons or other products. Their installation can introduce 'hidden' hazards.

Activated carbon bed canisters are known ignition sources under specific conditions. These conditions create a hot spot in the bed and possible ignition of a flammable-air mixture. A flashback into the storage tank causing an explosion can occur if the vapour space within the tank is within the flammable range. These factors and the need for preventive measures requires advice from the experts—consult your process safety management documentation for information and guidelines on this matter.

ACCIDENT **Explosion in internal floating roof tank!!!**

An explosion lifted a roof off of a storage tank, blowing it approximately 175 ft (50m). The cause of the explosion was an ignition of a flammable mixture in the vapour space of the tank (i.e. the space between the internal floating roof and the fixed roof) by an overheated carbon drum on the tank vent.

ACCIDENT **Wastewater separator explosion!!!**

An explosion occurred at the inlet division box of an API separator. Compliance with regulations some years ago to control VOC and benzene emissions required closure and sealing of these separators. Design changes included the installation of a nitrogen blanket and venting of the vapour space through carbon canisters to absorb hydrocarbon vapours.

The investigation identified three critical factors that caused the incident:

- a high concentration of volatile hydrocarbons in the separator;
- an overheated carbon canister that ignited a flammable mixture in the vent line and flashed back to the separator;
- a nitrogen blanket/purge that was insufficient to keep a low oxygen concentration in the separator.

Lessons learned

The introduction of carbon canisters to sewer and separator vents has resulted in safety-related risks that must be recognized and controlled.

HAZOPS of sewers, separator systems and tankage with carbon canisters should be revisited to ensure there are adequate precautionary measures in place to prevent fire and explosions.

> **Beware of hot spots in carbon canisters that could ignite a flammable atmosphere.**

9.8 Safe access to equipment

Safe access must be provided for personnel and, where applicable, vehicles at jetties. Below are some photographs depicting unsafe areas at terminals.

Anyone going to exposed areas of a jetty must wear a life jacket.

Missing handrail

Missing handrail

Unsafe access to jetty

No safety rail to prevent vehicles falling off jetty

ACCIDENT **Handrail severely corroded by corrosive environment!!!**

A wastewater treatment plant operator suffered fatal head and neck trauma falling from a stairway landing while performing routine work. Because there were no eyewitnesses, it is assumed that the operator either fell against (after slipping or tripping), or leaned against, or held onto a handrail section for fall protection at a stairway landing, and the two support legs, that were severely corroded, broke.

The operator fell 10 ft (3m) to the metal surface of a separator box. Investigation found that the 1947 vintage design was inadequate, as the number of repairs that have been made to railings and reported failures should have led to the identification of a systemic problem with civil structures. Facility handrails and walking/working surfaces were not part of an integrity inspection programme or civil structure management programme.

Heavy corrosive areas, such as jetties and some wastewater treatment plants, must be subjected to robust design and adequate inspection and maintenance.

9.9 Use of flexible hoses

Flexible hoses are used throughout refineries and terminals for various purposes where it is not practicable to have permanent hard piping. These temporary arrangements are often the last considered and least regulated components in either utility or process systems. They are expected to operate under a wide range of physical stresses and are therefore susceptible to failure.

Design and installation arrangements should be subject to a 'Management of Change' procedure and a HAZOP. Regular inspection and testing will be required to secure their continued fitness in service.

Good practice	Poor practice
1. Hoses on reels	
The outside diameter of the hose reel hub must be of greater value than the minimum bend diameter *(2x min bend radius, measured from inner wall of hose body)* of the hose to be applied.	*This steam hose assembly was installed on a reel 'hub' with a diameter 35% less than the minimum bend diameter of the hose.*
Reeled hose should be confirmed as being coiled to an acceptable bend radius.	Observation of severe bending of the hose at the hose-to-reel connection shall be cause for re-plumbing.
	The addition of adapters from the reel to the hose assembly should not cause over-bending immediately behind the hose fitting.
	Caution! Failure to consider these conditions may contribute to premature failure of the hose assembly.

continued

Good practice	Poor practice
2. Movement beyond design limits Planned movement of a hose assembly should be based on the 'line length' (exposed length of the hose body available for flexure) capabilities of the assembly, NOT the overall length.	 *This flexible metal hose assembly was installed in an application that demanded a lateral offset of 5 inches from centreline, twice its' capability. Bellows failure occurred at less than 100 cycles, shutting down a process.* Observation of extreme bends or deformations is evidence of movement beyond the capabilities of the hose assembly.
3. Horizontal hookups  Standpipes 12″ or more above ground level should be fitted with elbows with *outlets directed downward*. This will: • protect personnel from media stream in case of accidental disconnection or line rupture; • slow aging of the hose tube and oxidation of the reinforcement braid by 'straightening' the flow of steam; and • reduce bend stresses at the hose assembly end.	 **Caution!** Where observed, horizontal hookups for utility steam should be considered a risk to operators, and changed to vertical hookups.

Good practice	Poor practice
4. Bending stress Use of 45° or 90° elbows, bend restrictors, or hose slings are options available to correct this condition.	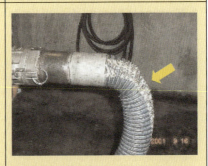 Severe downward bends observed in the hose body at the fitting juncture may be evidence of over-bending and signal impending failure.
5. Compatibility All hoses should be purchased to an engineering specification based upon the duty specified by the end users.	 Observation of pronounced hose body swelling and/or a 'spongy' feel indicates hose-to-media incompatibility, and is cause for immediate hose assembly retirement. **Caution! Steam Hose** The chlorobutyl tube in utility steam hose is not resistant to hydrocarbons. Contact with hydrocarbons will cause immediate and pronounced swelling of the tube, and shorten service life!

> **Ensure that hoses are used and inspected in accordance with the facility's safety procedures.**

Good practices

- Hoses should be colour-coded as far as it is practicable for example, red for steam, black for water, green for process.

- Different couplings should be used for different services to prevent any possibility of a mistake. All hoses should receive a pre-service inspection to ensure that they meet specification and are not damaged.

- All hoses should be examined by the operator immediately prior to use.

- All hoses shall receive a periodic examination and/or test by the 'Inspection Authority'.

- A HAZOP should confirm that when a hose is used in a closed circuit the operating pressure and temperature of the system cannot exceed the design limits of the hose.

- Disconnecting of a flexible hose must be treated in the same way as a pipeline disconnection.

- All hoses used for transfer of flammable materials must be electrically continuous to prevent the accumulation of static electricity.

- All hoses should be registered with a specific number on the Inspection Record.

- Defective hoses shall be immediately destroyed to ensure that they cannot be inadvertently used prior to disposal.

ACCIDENT **Major oil spill due to failure of compression joint!!!**

A major oil spill (40 tonnes) occurred during the transfer of off-specification lube oil back to the refinery from a jetty storage tank using a temporary diesel driven pump connecting permanent pipelines with flexible hoses.

After several hours of pumping, the pump's discharge hose compression fitting located at the flanged connection with the permanent piping failed. The hose was supplied by a third party contractor and found to have a fabrication defect. Although the end fitting had been swaged into the hose, this was not adequate. The manufacturer re-ended the failed fitting using a tighter ferrule and larger swaging dolly.

Close up of failed compression joint

9.10 Health risks

A Material Safety Data Sheet (MSDS) must be available to all employees for every product that is handled at the facility.

The information provided in the MSDS provides the basis for assessing the risks to health from all activities carried out at the site. Additional information should be obtained from the manufacturer.

Sufficient controls must be in place to ensure that employees and others are not exposed to levels of the substance above those stipulated in national standards.

Personal Protective Equipment for a specific product/chemical should be considered a temporary measure and may include respiratory protection, protective clothing, footwear and eye protection. Permanent engineering controls should always be installed if possible to ensure minimal exposure to hazardous materials.

Emergency procedures must cover the immediate actions to be taken should there be an uncontrolled release of a hazardous material and/or an overexposure exceeding acceptable limits.

Refer to BP Process Safety Booklet *Hazardous Substances in Refineries* for further details.

10

Some points to remember

1. Flammable atmospheres should be avoided in fixed roof tanks.

2. The liquid's true vapour pressure is usually the main determining factor for the selection of type of storage tank.

3. The flammable range is the concentration range lying between the lower and upper flammability limits (LFL/LEL and UFL/UEL).

FLAMMABLE ATMOSPHERE

4. Mists can form flammable mixtures below their flashpoints.

5. The bigger the fixed/cone roof storage tank, the more vulnerable it is to failure from overpressure!

6. Ensure that vents, flame arrestors and other mesh screens are regularly inspected to make sure they are clear of rags, rust, ice, tissues, wax, polymer, plastic and other debris.

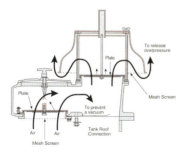

7. Refer to *API Standard 2000* for further details on causes of overpressure and vacuum, venting requirements, and installation and maintenance of venting devices.

8. Protect against the inhalation of vapours which escapes as a momentary 'whoosh' when a gauge hatch is opened on a fixed/cone roof tank equipped with a PV valve.

9. Keep potential sources of ignition away from storage tanks.

10. A bund/dike should be capable of containing at least the capacity of the largest tank within the bund/dike.

11. Bund/dike drain valves must be kept closed except when the bund/dike is being drained of rainwater.

12. Incompatible chemicals should not be stored in tanks within the same bund/dike.

13. Valves on roof drains should be kept open at all times.

14. A tank with a leaking internal drain should be taken out of service as soon as practicable for repair.

15. Violent foaming action caused by the vaporization of water will result in a 'froth-over'. Keep the contents of tanks below 93°C (200°F) to prevent this occurrence.

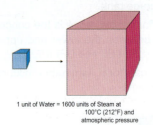

1 unit of Water = 1600 units of Steam at 100°C (212°F) and atmospheric pressure

16. When did you last perform a HAZOP on your slops systems?

17. Never leave a tank unattended when draining off water.

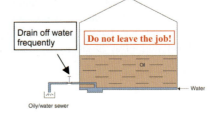

Drain off water frequently

Do not leave the job!

Oil

Water

Oily/water sewer

18. Electrostatic accumulation in liquids is significant *unless**:

- conductivity exceeds 50 pS/m (resistivity less than 2×10^{12} ohm cm), and

- chemical is handled in earthed/grounded conductive containers.

*Not applicable for mists

< 50 pS/m?

19. Larger electrostatic charges are obtained with:
    - filters;
    - pumps;
    - lower conductivity liquids;
    - smaller diameter pipes;
    - presence of water and particulate matter;
    - splash filling;
    - increased velocity/flowrate.

20. Static Electricity Discharge
    + Flammable Atmosphere
    = Danger of Fire and Explosion

21. Earth/ground and bond all equipment and containers to prevent static charge build-up.

22. Remember! You cannot see a static charge build-up. It is too late when a spark is produced. Observe your tank operating procedures and report all incidents of static electricity.

23. Switch loading is a potentially hazardous operation.

24. Do not use mobile phones where they are prohibited. If they are allowed, avoid being distracted and focus on the task at hand!!

25. Remember there are many potential sources of ignition that must be strictly controlled including diesel driven equipment.

26. For high risk products (in the flammable range and low conductivity <50 pS/m) follow strict procedures for loading, gauging and sampling.

27. Do not modify procedures or equipment without authorization through the Management of Change procedure.

28. Do not splash fill (mists can form flammable mixture below flash point).

29. Do not remove loading pipes or take samples from containers immediately after completion of filling. Wait the required relaxation time.

30. Beware of filters, water, line blowing that increases static build-up.

31. Elimination of flammable mixture (such as through inert atmosphere) removes the static problem.

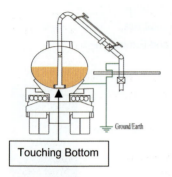

Touching Bottom

32. Ensure the end pipe configuration (top loading) touches the bottom of the compartment to avoid splash filling.

33. In the case of bottom loading, unless there is an earthed conducting rod from the top to the bottom of the compartment, giving an electrical effect similar to the fill pipe, there is an increased possibility of electrostatic discharge and so the loading velocities are reduced.

34. THE FIRST TASK:

Earth/ground and bond tanker/road truck, equipment and containers to prevent static charge build-up.

Earthing/Grounding

35. Loading activities that can cause a static electricity hazard:
- splash filling;
- switch loading;
- sampling, dipping and taking temperatures.

36. Loading bays should be equipped with alternative escape routes with emergency stop buttons and an additional button located remote from the rack (approximately 30m (98 ft) away).

37. Isolation valves should be located where they can be easily and safely reached under fire conditions or preferably each product line should be equipped with an emergency isolation valve that can be operated by a single switch from a remote location.

38. The emergency system should be designed to 'fail safe', i.e. valves close in the event of air or power failure.

39. Ensure road trucks ready to load are parked at a safe distance from the gantry and their movement is strictly controlled by the supervisor.

40. Wear the correct Personal Protective Equipment when handling hazardous products and chemicals.

41. Check that the right product will go into the right tank before unloading commences through sampling, documentation and labelling.

42. Does the terminal have a procedure in place to ensure that a pre-cargo transfer conference is undertaken, including completion of the ship/shore safety checklist?

43. A ship/shore bonding cable is not effective as a safety device and may even be dangerous. A ship/shore bonding cable should therefore not be used.

44. Use an insulating flange or a single length of non-conducting hose to ensure electrical discontinuity between ship and shore, as required by ISGOTT.

45. Stray currents and static electricity are different hazards that require different preventative measures.

46. The magnitude of the pressure surge depends on the closure time of the valve that in turn depends upon the design of the valve.

47. Emergency Release Coupling must only be released if the isolating ball valves are shut first. The integrity of the interlock mechanism must be guaranteed.

48. Inert gas systems must remain fully operational during the discharge of cargo from ships.

49. Critical items associated with the safe operation of storage tanks shall be formally checked by the Operator on a regular basis.

50. Special notes!
   - When the floating roof is in a high position the opportunity should be taken to check the following:
     o drain sump;
     o pontoon compartments;
     o seal and weathershields;
     o emergency roof drain;
     o rolling ladder;
     o metallic shunts (lightning protection).

   Such inspections should be authorized through a work permit.
   - Binoculars can be used to good effect for viewing details on floating roofs.
   - The tank and its roof must be in a safe condition (for example, no explosive atmosphere in vapour space and roof not too thin for access) before such inspections are carried out.
   - A standby attendant, to raise the alarm in case of an emergency, is required before operators go onto a floating roof for inspection.
   - Access onto floating roofs should preferably only be authorized at a high level (5 ft/1.5m) from the top. A confined space entry permit will be required if the tanks contain toxic materials, such as H_2S, or if the roof is at a lower level.
   - Beware of H_2S hazard from certain crude oil tanks. Breathing apparatus may be required to be worn before accessing the tank's floating roof.
   - Safe entry upon floating roofs must be taken seriously. Refer to API Publication 2026 for details on the general precautions and potential hazards associated with entry onto floating roofs, as well as procedures to be followed prior to entry.

51. The maximum period between internal inspection of a storage tank is 20 years if the corrosion rate is known and not found to be critical, or ten years if the corrosion rate is unknown.

52. The removal of a storage tank from service to recommissioning may take many months. Each step in the process must be authorized through the work permit system and the various tasks subject to a Job Hazard Analysis to assure the adequacy of the proposed safeguards.

53. Beware of pyrophorics! They may generate heat and ignite causing explosion and fire. It is important to keep any pyrophoric scale wet until it can be removed to a safe location.

54. A Risk Assessment needs to be done prior to entry into a tank for gas testing, and controlled through a Confined Space Entry Permit in accordance with the site's Safety Rules.

    Gas tests and a visual inspection will be required to determine whether there is a safe working environment for hot work.

55. Standby attendants are strictly not allowed to enter the tank under any circumstances!

56. Local exhaust ventilation is required for welding work in a confined space.

57. Anyone going to exposed areas of a jetty or a wastewater treatment plant must wear a life jacket.

58. Check flanges, valve glands and packings for fugitive emissions.

59. Protect the environment by minimizing hydrocarbon liquid evaporation losses to the atmosphere, for example, by keeping equipment in good order, repairing leaks immediately, keeping sample/dip hatches closed.

60. Beware of hot spots in carbon canisters that could ignite a flammable atmosphere.

61. Ensure that hoses are used and inspected in accordance with the facility's safety procedures.

Test yourself!

General

1. The liquid product's vapour pressure is usually the main determining factor in selecting the type of storage tank.

    True ☐   **False** ☐

2. Any petroleum product with a TVP greater than 11.1 psia (0.77 bara) should be stored in floating roof tank.

    True ☐   **False** ☐

3. It is alright to store petroleum products with flash point below 55°C (131°F) in fixed roof tanks.

    True ☐   **False** ☐

4. Products with low flash points can easily form a flammable atmosphere with air at ambient temperature.

    True ☐   **False** ☐

5. Floating roof tanks are used to store products with low flash points because it eliminates flammable vapour headspace.

    True ☐   **False** ☐

6. In a floating roof tank, the roof sits on top of the product liquid surface and floats up and down as the liquid level fluctuates during loading and unloading.

    True ☐   **False** ☐

Tank and bunding/diking

7. One of the most common causes of failure of a fixed roof tank is the vent being choked, blanked off, wrapped with a plastic bag, modified or inadequately sized.

    True ☐   **False** ☐

8. It is alright to locate several tanks together less than 1m (3.3 ft) apart.

    True ☐   **False** ☐

9. It is alright to build a new tank within a small bund/dike with other existing tanks.

    True ☐   **False** ☐

10. It is alright to locate both alkali and acid tanks within the same bund/dike.

    True ☐   **False** ☐

11. When the internal drain hose of a floating roof tank is leaking, it is adequate to shut off the roof drain valves and continue operating the tank because the floating roof will not sink during heavy rainfall.

    True ☐   **False** ☐

12. Froth-over is a violent foaming action caused by the rapid vaporization of water.

    True ☐   **False** ☐

13. It is alright to leave a tank draining job unattended.

    True ☐   **False** ☐

14. Bund/dike drain valves must be kept shut to collect any spilled or leaking product, except when draining rainwater.

    True ☐   **False** ☐

15. Internal inspection of tanks must be undertaken at least every year.

    True ☐   **False** ☐

Electrostatic

16. Electrostatic build-up and discharge (source of ignition) is a major hazard when handling products or chemicals with a conductivity less than 50 pS/m.

    True ☐   **False** ☐

17. An electrostatic discharge in a flammable atmosphere can cause fire and even explosion.

    True ☐   **False** ☐

18. Pumps, small pipes, filters, fittings and valves are common electrostatic generators in piping systems.

    True ☐   **False** ☐

19. Agitation, mixing, splash filling, switch loading, steaming and grit blasting are some operations that can generate static electricity in the process plants.

    True ☐   **False** ☐

20. Electrostatic accumulation can be prevented by the use of earthing and bonding, anti-static additives or relaxation time.

    True ☐   **False** ☐

21. Plant personnel should use anti-static footwear to guard against a static hazard.

    True ☐   **False** ☐

175

Loading/unloading

22. A product can form a flammable atmosphere even below its flash point when it exists in mist form (fine liquid droplets suspended in air) generated by splash filling.

    True ☐   **False** ☐

23. A stilling pipe or dip pipe that touches the bottom of the storage tank should be used for sampling or dipping to assist the dissipation of a static charge.

    True ☐   **False** ☐

24. The first task during loading and unloading of a road/rail tanker is to earth/ground and bond road/rail tanker, loading pipes, hoses and container to dissipate a static charge build-up.

    True ☐   **False** ☐

25. For top loading of road tanker, ensure the end pipe touches the bottom of the compartment to avoid splash filling.

    True ☐   **False** ☐

26. For bottom loading, the loading velocities are reduced (as compared to top loading) as there is an increased accumulation of an electrostatic charge in the absence of a fill pipe.

    True ☐   **False** ☐

27. It is not necessary to install emergency stop buttons at loading bays.

    True ☐   **False** ☐

28. The emergency isolation valves should be designed to fail open in the event of air or power failure.

    True ☐   **False** ☐

29. Each product line should be equipped with an emergency isolation valve at a loading gantry that can be operated by a single switch from a remote location.

    True ☐   **False** ☐

30. It is alright for the road/rail tanker personnel to tamper with the spring-loaded nozzle (deadman's handle) so that he can be temporarily relieved from loading/unloading duties.

    True ☐   **False** ☐

31. Excessive vapour emission during loading/unloading can be effectively reduced by a vapour return line.

    True ☐   **False** ☐

32. It is not necessary to conduct a pre-cargo transfer conference or complete the ship/shore safety checklist prior to loading/unloading to/from a ship.

        True ☐   **False** ☐

33. It is alright to start loading the ship at the maximum flow rate immediately so that the loading can be completed faster.

        True ☐   **False** ☐

34. It is alright to change the loading rate substantially without informing the ship.

        True ☐   **False** ☐

35. Stray currents between ship and shore can generate an incendive arc (a source of ignition) when it is suddenly interrupted during disconnecting of the loading hose or arm at the ship's manifold.

        True ☐   **False** ☐

36. The use of insulating flange or a single length of non-conducting hose can provide protection against stray currents.

        True ☐   **False** ☐

37. Insulating flanges should be painted and greased regularly.

        True ☐   **False** ☐

38. Emergency release coupling (ERC) should only be released if the isolating ball valves are shut first.

        True ☐   **False** ☐

39. Booms and skimmers can be used to contain and remove oil slicks in the event of accidental oil spillage.

        True ☐   **False** ☐

40. To avoid surge pressures when loading a ship, use the correct emergency shutdown sequence, i.e. first shut down the terminal, then the ship (always shut down the source of supply first, followed by the receiving party).

        True ☐   **False** ☐

41. To avoid surge pressures, reduce loading/loading rates to a safe level corresponding to the type of isolation valve used.

        True ☐   **False** ☐

42. It is usual to provide an independent high-high level alarm for storage tanks to protect against overfilling.

        True ☐   **False** ☐

43. It is necessary to monitor the tank level continuously during loading and unloading.

        True ☐   **False** ☐

Switch loading

44. The most significant hazard occurs when switch loading a low volatile low conductivity product into a compartment of a road truck that had previously carried a low flash point product such as gasoline.

        True ☐   **False** ☐

45. It is alright to load a low flash point product into any tank without first checking the previous cargo held by the tank.

        True ☐   **False** ☐

46. Switch loading can arise from sharing a common manifold/pipeline for various products.

        True ☐   **False** ☐

47. Purge with nitrogen or gas-free the vessel prior to switch loading.

        True ☐   **False** ☐

Miscellaneous

48. It is not necessary to dowse pyrophoric scale with water and keep it wet at all times.

        True ☐   **False** ☐

49. If the oxygen meter reads 18%, the contaminant level in the air being tested is 13.5%.

        True ☐   **False** ☐

50. Local exhaust ventilation is not required for welding work in a confined space.

        True ☐   **False** ☐

46T/47F/48F/49T/50F
39T/40T/41F/42T/43T/44T/45F
32F/33F/34F/35T/36T/37F/38T
25T/26T/27F/28F/29T/30F/31T
18T/19T/20T/21T/22T/23T/24T
11F/12T/13F/14T/15F/16T/17T
1T/2F/3F/4T/5T/6T/7T/8F/9F/10F
ANSWERS

Bibliography

ANSI B16.5: Pipe Flanges & Flanged Fittings

ANSI B31: ASME Process Piping

API 2220: Improving Owner And Contractor Safety Performance

API Publication 581: RBI (Risk-Based Inspection)

API Publication 2026: Safe Access/Egress Involving Floating Roofs Of Storage Tanks In Petroleum Service

API RP 2003: Protection Against Ignitions Arising Out Of Static, Lightning, And Stray Currents

API Standard 2015: Safe Entry and Cleaning of Petroleum Storage Tanks

API RP 2016: Guidelines and Procedures for Entering and Cleaning Petroleum Storage Tanks

API RP 2350: Overfill Protection For Storage Tanks In Petroleum Facilities

API Standard 2000: Venting Atmospheric And Low-Pressure Storage Tanks—Nonrefrigerated And Refrigerated

API Standard 2205: Guide For The Safe Storage And Loading Of Heavy Oil And Asphalt

API Standard 2610: Design, Construction, Operation, Maintenance and Inspection of Terminal and Tank Facilities

API Standard 650: Welded Steel Tanks For Oil Storage

API Standard 653: Tank Inspection, Repair, Alteration, And Reconstruction

API 2021: Management of Atmospheric Storage Tank Fires

BS 2654: Specification For Manufacture Of Vertical Steel Welded Non-Refrigerated Storage Tanks With Butt-Welded Shells For The Petroleum Industry

EEMUA Pub No 159: Users' Guide To The Maintenance And Inspection Of Above-Ground Vertical Cylindrical Steel Storage Tanks

IEC.61508: Functional Safety Of Electrical/Electronic/Programmable Electronic Safety-Related Systems

IP: Institute of Petroleum Model Code of Safe Practice in the Petroleum Industry, Part 3, Refining

ISGOTT: International Safety Guide for Oil Tankers and Terminals

NFPA 30: Flammable And Combustible Liquids Code

NFPA 51: Design and Installation of Oxygen-Fuel Gas Systems for Welding, Cutting, and Allied Processes

NFPA 70: National Electric Code

NFPA 77: Static Electricity

OCIMF: Oil Companies International Marine Forum

Guide For Handling, Storage, Inspection and Testing of Hoses in the Field, OCIMF.

Design and Construction Specification for Marine Loading Arms, OCIMF.

Loss Prevention in the Process Industries: Hazard Identification, Assessment and Control, Frank P. Lees.

Liquefied Gas Handling Principles on Ships and in Terminals, McGuire and White.

Liquefied Gas Fire Hazard Management, SIGTTO, June 2004 ISBN 1 85609 265 7.